Integrated Circuit Projects

Integrated Circuit Projects

Written By:
Carl J. Bergquist

PUBLICATIONS

A Division of Howard W. Sams & Company
A Bell Atlantic Company

©1997 by Howard W. Sams & Company

PROMPT© Publications is an imprint of Howard W. Sams & Company, A Bell Atlantic Company, 2647 Waterfront Parkway, E. Dr., Indianapolis, IN 46214-2041.

All rights reserved. No part of this book shall be reproduced, stored in a retrieval system, or transmitted by any means, electronic, mechanical, photocopying, recording, or otherwise, without written permission from the publisher. No patent liability is assumed with respect to the use of the information contained herein. While every precaution has been taken in the preparation of this book, the author, the publisher or seller assumes no responsibility for errors or omissions. Neither is any liability assumed for damages resulting from the use of information contained herein.

International Standard Book Number: 0-7906-1116-3
Library of Congress Catalog Card Number: 97-65786

Acquisitions Editor: Candace M. Hall
Editor: Loretta L. Leisure
Assistant Editors: Pat Brady, Natalie F. Harris
Typesetting: Loretta L. Leisure
Indexing: Loretta L. Leisure
Cover Design: Phil Velikan
Graphics Conversion: Terry Varvel, Walt Stricker
Illustrations and Other Materials: Courtesy of the Author

Trademark Acknowledgments:
All product illustrations, product names and logos are trademarks of their respective manufacturers. All terms in this book that are known or suspected to be trademarks or services have been appropriately capitalized. PROMPT© Publications, Howard W. Sams & Company, and Bell Atlantic cannot attest to the accuracy of this information. Use of an illustration, term or logo in this book should not be regarded as affecting the validity of any trademark or service mark.

PRINTED IN THE UNITED STATES OF AMERICA

9 8 7 6 5 4 3 2 1

Table of Contents

Table of Contents	5
Introduction	9
Project #1:	
Stereo Bargraph VU Meter	13
Project #2:	
DTMF Decoder	19
Project #3:	
ISD1000A Voice Recorder	27
Project #4:	
Digital Audio Frequency Meter	35
Project #5:	
Temperature Detector	43
Project #6:	
TTL Counter	49
Project #7:	
Up/Down Counter	57
Project #8:	
ICM7217 Digital Stopwatch	63
Project #9:	
Digital Alarm Clock	71

Integrated Circuit Projects

Project #10:
  4017 Counter/Divider   79
Project #11:
  FM Wireless Microphone   89
Project #12:
  RF Bug Detector   99
Project #13:
  Passive Detector Receiver   107
Project #14:
  RCA Voltmeter   113
Project #15:
  7107 Digital Thermometer   119
Project #16:
  8038 Function Generator   127
Project #17:
  Logic Probe   133
Project #18:
  Color Organ   141
Project #19:
  Laser Diode Driver Circuits   147
Project #20:
  Audio Tone Generator   155
Project #21:
  TV Modulator   161
Project #22:
  Touch-Tone Generator   167
Project #23:
  Frequency to Voltage Converter   173

Table of Contents

Project #24:
  Infrared Transmitters and Receivers   179
Project #25:
  ZN414Z Based AM Radio   191
Conclusion   197
Appendix A   201
Index   207

Integrated Circuit Projects

Introduction

This book came about as the result of a conversation with Candace Drake. For those of you who don't know Candace, she is the lady I deal with at Prompt® Publications (her official title is Managing Editor, but we don't want it going to her head).

Anyway, a while back I asked Candace what kind of books she was interested in, and she recommended a project book for integrated circuits. You know, the type you can breadboard quickly, and see if you like it. It sounded good to me, and so, *IC Projects* was born, or perhaps more accurately, conceived.

In reality, the text's creation wasn't as simple. In preparation, I thought about the many books regarding this subject I had read and used over the years, and decided the best approach was to write a volume containing the projects I would like to see. My "Dream Book", if you will.

Hopefully, the circuits presented would also be what other experimenters were looking for. This did afford a challenge, however, as the variety of ICs available is probably surpassed only by the variety of Asian Flu strains.

So, some practical parameters had to be established. The first was (drum roll, please) no bizarre, weird, impossible to find

Integrated Circuit Projects

without sending off to Istanbul, or some equally outrageous destination ICs. I have made every effort to provide designs that use commonly available ICs (at least as of this writing).

I don't know about you, but few things frustrate me more than to read a construction piece that utilizes some IC I can't find in any of the catalogs, or is made by a company I've never heard of (i.e. The Ma Jones Apple Pie, Inner Tube and Integrated Circuit Company). I invariably end up tossing the article in disgust, but worse, it's a waste of time, effort, and publication space.

Thus, you WILL be able to find the chips in this book. On some occasions, I will include certain suppliers in the parts list, but the IC WILL be available. If not, write me an ugly letter.

Another parameter is cost. There have, over the years, been countless projects that held great interest for me, but I sure hated the part about having to re-finance my car to pay for them. Again, every effort has been made to keep the expense within "hobby" range.

A third aspect, regarding the ICs, is ease of use. For example, and I am not trying to knock microcontrollers, not at all, they serve a very useful purpose, but how many of us have, or have access to, a programmer for the Motorola series, and a programmer for the Intel series, and a programmer for the PIC series? You see what I mean? We are back to expenses many hobbyists can't afford.

There were some other parameters, but I'm not going to get into those. This is probably beginning to bore you. Let me just say that this text is intended to be, as much as I hate to use this tired, overworked phrase, user friendly.

Nearly all the projects employ integrated circuits. The FM transmitter and the laser diode driver use transistors, but after all, what is an integrated circuit? Just a bunch of transistors. They run the gamut from a handful of gates with 20 semiconductors to sophisticated microprocessors containing 20,000,000 transistors or more. So, I reckon you can forgive me a couple of designs.

OK, once again, here it is; *IC Projects*, Volume 1 (whether there is a volume two depends largely on you, the consumer). I do hope you like it, and I think you will. Candace liked it. She must have, otherwise you wouldn't be reading this.

Have fun, and THANKS!

Carl

Integrated Circuit Projects

What You Will Need:

- ✔ U1/U2 — LM3916 VU Meter Display Drivers
- ✔ LED1-20 — Jumbo T-1 3/4 Light Emitting Diodes (any color)
- ✔ D1 — 1N4001 Rectifier Diode
- ✔ R1/R2 — 1,000 Ohm, 1/4 Watt Carbon Film Resistors
- ✔ R3/R4 — 5,000 Ohm Potentiometers
- ✔ T1/T2 — 8 to 2,000 Ohm Audio Transformer (Jameco)
- ✔ J1/J2 — Audio Phone, or RCA Jacks
- ✔ S1/S2 — SPDT Slide or Toggle Switches
- ✔ S3 — DPDT Slide or Toggle Switch

Note:

Connection to the audio source can be done by phone plugs, RCA style plugs, or any well shielded system. It is advisable to use coaxial cable for the lead wires, as this prevents line loss and noise.

Integrated Circuit Projects

Appendix A provides a list of parts suppliers that have proven reliable in the past. If a component tends to be difficult to locate, or is special purposed, one possible source is listed in the parts list, for convenience. That is not to say that other sources won't work as well.

One highly aesthetic addition to any project involving audio is the 10 step LED VU meter. Such projects can include stereo equipment, receivers, communications transceivers, PA amplifiers, tape decks, etc. Anything that can drive a speaker can incorporate this LED meter as an indicator.

The goal of the circuit is to provide a visual representation of the audio stage output intensity (loudness). While analog meters can be employed, the rise and fall of a row of LED's provides a more pleasant, and much easier to read display.

Instead of having to zero in on that moving needle, usually just a glance at the LED VU will indicate the level. As for stereo, a dual meter, like the one illustrated in this chapter, makes for very simple balancing of the separate left and right channels.

Aside from the practical applications, these meters contribute a futuristic appearance to any front panel. This is not to say that there are not times when an analog meter is necessary, or preferable, only that it is prudent to be familiar with as much technology as possible. In that fashion, the most pragmatic and appealing methods are both at your disposal.

Project #1: Stereo Bargraph VU Meter

So much for the sales pitch. The magic of a single chip allows a normally complicated piece of hardware to be constructed simply and quickly. So, let's look at this circuit.

CIRCUIT THEORY

This special chip is the linear LM3916 VU Meter Display Driver, and is one of three similar ICs. In addition to the 3916, the series includes the LM3914 and LM3915, with the 3914 acting as a bargraph driver, while the 3915 provides steps in degrees of 3dB each. Pinouts on all three chips are identical.

Originally designed by National Semiconductor, the LM3914 series offers compact size, a wide range of operating voltage, and excellent versatility. All of this has led to the chip's becoming an Industry Standard.

The principle behind this IC involves both a voltage divider and a comparator circuit. A changing voltage is applied to the comparator's inverting input (−), while the non-inverting input receives the product of a voltage divider with ground on one end, and the supply voltage on the other (aka: reference voltage).

When the input voltage reaches, or exceeds the reference voltage, the comparator's output goes low, thus lighting the LED connected to that output. Additionally, pin 9 of the LM3916 allows for the selection of either a single illuminated LED (dot mode), or a series of LEDs progressively lighting as the voltage rises to each individual reference voltage (bar mode).

The bar mode is more often used as it provides the lighted row of LEDs effect, but the dot mode can be very handy in certain situations. Normally required LED dropping resistors are not needed, as this function is built into the LM3916.

Integrated Circuit Projects

When the audio signal's intensity changes, so changes the input voltage, and the number of LEDs that light (bar mode). In the case of the dot mode, the position of the glowing LED changes in response to the input voltage.

CIRCUIT DESCRIPTION

Looking at the schematic diagram, **Figure 1-1**, it is seen that this configuration is a dual chip stereo style meter. However, it is an easy matter to modify the system. If you only need one ten step meter, simply cut the circuit in half, as both sides are virtually the same. On the other hand, if you want a twenty step meter, U1 and U2 can be cascaded by tapping in at the U1 pin 10/LED connection, and running a line to U2's pin 5.

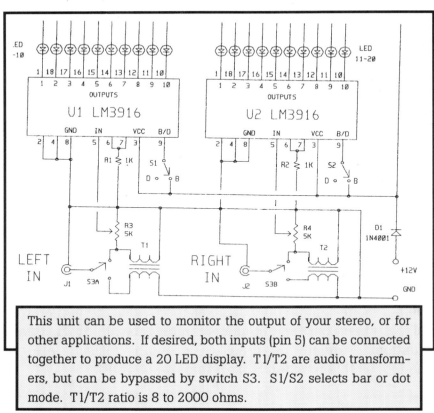

This unit can be used to monitor the output of your stereo, or for other applications. If desired, both inputs (pin 5) can be connected together to produce a 20 LED display. T1/T2 are audio transformers, but can be bypassed by switch S3. S1/S2 selects bar or dot mode. T1/T2 ratio is 8 to 2000 ohms.

Figure 1-1. Dual VU meter schematic.

Project #1: Stereo Bargraph VU Meter

So, **Figure 1-1** is as much a guideline as it is a rigid schematic. As stated before, the two sections are duplicates of one another which makes the configuration simple to follow. The only common lines come from the power supply.

In the way of a more in-depth explanation, let's ponder U1's circuit. The input (amplifier output) is applied at jack J1, and can be directed, via switch S3a, to either an audio transformer T1, or directly to potentiometer R3. The center tap, or whipper of R3 is connected to U1's pin 5, which is the IC's input.

The LM3916 outputs are found at pins 1, and 10-18, with the LED cathodes going to the IC, and anodes tied to the positive supply rail. Pin 9 of both ICs provides the dot/bar mode selection. When hooked to the positive line, the bar, or row of lights effect is achieved. If connected to ground, or left floating, the single LED display appears.

Mode selection is controlled by switches S1 and S2, and, of course is a matter of preference. R3 acts as a sensitivity control so that the meter can accommodate a wide range of output powers, and T1 provides matching for an 8 ohm output, often found on audio equipment.

Ultimately, thanks to the LM3916, this project is not at all complicated. You can breadboard it first to get the feel of the circuit, then install it in equipment or its own case, if you decide to put the meter to use. Either way, it makes an interesting and entertaining project.

Regarding construction methods, the circuit's simplicity lends itself to point-to-point wiring, or if you are equipped, a PCB design. Both approaches work equally well. A PCB will provide a slightly more durable finished product, but unless the board is not to be enclosed, durability shouldn't be a problem.

Integrated Circuit Projects

CONCLUSION

While I have set up this circuit as a stereo VU meter, the chip can be utilized in a number of other areas. For example, it can act as a silent output for a "bug detector". In that fashion, the clandestine listening device can be located without tipping off the listener.

The LM3914 provides the basis for an excellent RF power/signal meter for communications gear. This is especially true with portable equipment, as the device is far less fragile than standard analog meters. overall, it will usually take up less space, allowing for a more compact design.

Rigged with a small amplifier and microphone, it can be a great addition to a child's toy. In this application, the LEDs will respond to changes in sound, or the child's voice. Keeps 'em busy for hours.

So, with a little imagination, there's no telling where you can go with this thing. Virtually any source of changing voltage will act as an input for the meter (as long as the voltage doesn't exceed the operational voltage of the circuit), and the results can be calibrated to indicate useful information.

All in all, the LM3914 series represents another in a long line of highly versatile, highly pragmatic devices. I can't help but wonder sometimes if the Bell Laboratory engineers who invented the integrated circuit ever envisioned the magnitude IC users' ingenuity would reach.

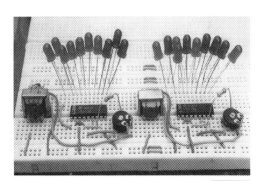

Photo 1-1. A solderless breadboarded version of the dual channel LED VU bargraph.

What You Will Need:

- ✔ U1 — 75T202 Low Power DTMF Receiver
- ✔ U2 — 74LS47 BCD to 7 Segment Decoder/Driver
- ✔ V1 — V130LA2 Metal Oxide Varistor (JDR)
- ✔ LED1 — Green Jumbo T-1 3/4 Light Emitting Diode
- ✔ DSP1 — Red 0.5 Inch 7 Segment LED Display
- ✔ ZENER — 1N4733 5.1 Volt Zener Diode (optional - see text)
- ✔ R1 — 1,000,000 Ohm, 1/4 Watt Carbon Film Resistor
- ✔ R2 — 220 Ohm, 1/4 Watt Carbon Film Resistor
- ✔ R3 — 470 Ohm, 1/4 Watt Carbon Film Resistor

Note:

Some arrangement needs to be made for hook-up to the telephone line. The easiest way is to use a length of cable that already has a modular plug on one end. Attach the RING (red lead) and TIP (green lead) as illustrated in **Figure 2-2**.

Integrated Circuit Projects

Whenever a component is a special purpose, or hard to find item, one possible supplier is listed after that component in the parts list. These are, of course, not the only sources, and you may be able to obtain the part elsewhere.

When you hear the tone pair generated by a touch-tone telephone, have you ever wondered what number was represented? If so, this is a project for you. In addition to telephones, these tones are used to trigger amateur radio repeaters, can be heard on scanners, radio and television programs, and often are used for remote control purposes.

Thus, there are many times when it is important, or at least amusing, to know what the tones denote. Through the use of two commonly available ICs, and a seven segment LED display, a device can be constructed that will provide you with just that information.

CIRCUIT THEORY

The principle behind this circuit is to first decode the tone pair, then display the digit it represents. If not for a crystal controlled decoder IC, this would lead to a complicated array of individual tone and binary decoder chips, hardly the simplistic configuration employed in this design. However, we have such a chip, and it functions as the brains in the following description.

Project #2: DTMF Decoder

The first section of the circuit involves the input signal, which can be either a straight audio input, as from a receiver, or directly from the telephone line. If the phone line is the source, the input to the tone decoder has to be protected and isolated from the line, using a varistor and transformer.

More will be said about this later, but the primary problem is the wide variation of voltages on a telephone line. These can run as high as 100 volts (phone ringing), and since the IC decoder uses a 5 volt supply, 100 volts would fry it.

Once the input signal is conditioned, it is introduced to the input of the tone decoder chip. Here, the twin tones are detected, decoded, and the appropriate BCD code is assigned to the IC's four pin output. Additionally, the decoder determines if the tone pair is, indeed, a valid signal, or just noise on the line. If valid, the Data Valid (DV) pin goes high, and this can be used to activate an LED.

The tone decoder output consists of four individual signals, or bits (this is know to the computer folks as a nybble), and is applied to the second IC, a BCD to 7 segment decoder/driver. This chip converts the binary code to an output that lights the appropriate segments of the display.

Everybody got that? Well, maybe not, so let's go through it again. First, however, refer to **Figure 2-1**, as this should provide an understanding of how the 7 segment display is set up. As can be seen, the unit consists of seven bar shaped elements designated a, b, c, d, e f, and g. If the number "5" is to be represented, then the a, c, d, f, and g bars, or segments, must be illuminated. Elements b, and e remain dark.

OK, back to our tale. When you press the "5" button on a telephone, the tone pair of 770 hertz and 1336 hertz (**Table 2-1** provides all combinations) is generated and sent to the tone decoder IC, U1. U1 then detects these tones and issues a binary output equivalent to decimal "5" (BCD for 5 is "0101").

Integrated Circuit Projects

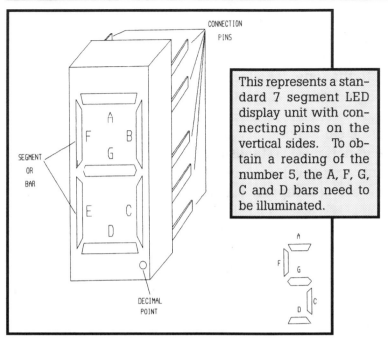

Figure 2-1. LED display unit segment designations

TELEPHONE KEYS	LOW FREQUENCY	HIGH FREQUENCY
0	941 HZ	1336 HZ
1	697 HZ	1209 HZ
2	697 HZ	1336 HZ
3	697 HZ	1477 HZ
4	770 HZ	1209 HZ
5	770 HZ	1336 HZ
6	770 HZ	1477 HZ
7	852 HZ	1209 HZ
8	852 HZ	1336 HZ
9	852 HZ	1477 HZ
*	941 HZ	1209 HZ
#	941 HZ	1477 HZ

Table 2-1. Frequency assignments for telephone keypad.

Project #2: DTMF Decoder

This code is now sent to the second IC, again decoded, the display is told to turn on segments a, c, d, f, and g. That combination, of course, appears to the eye as the number "5", and if all is well, the tone decoder indicates so by sending its data valid pin high.

Now, isn't that better? It is for me. In essence, each time a tone pair is introduced to this circuit, the display will show the corresponding number. One point to remember, though. The display only lasts as long as the tone pair is present (the telephone key is pressed), so sometimes you have to be quick to get all the numbers.

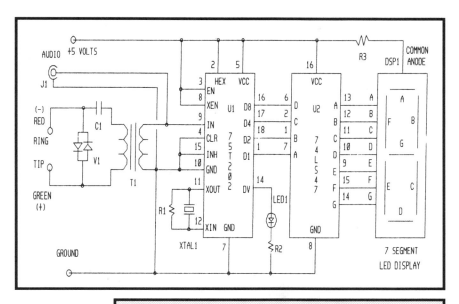

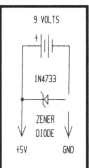

This device will allow you to see each number as it is entered by the telephone. The unit can be connected to the telephone line at the ringer (red), and tip (green) wires, or to an audio source through the jack J1. The latter will allow the DTMF tone pairs to be taken from the output of a receiver.

This circuit will provide the +5V regulated power supply.

Figure 2-2. Single digit DTMF decoder schematic.

CIRCUIT DESCRIPTION

With theory out of the way, let's look at the nuts and bolts of this project. Referring to the schematic in **Figure 2-2**, it is seen that the circuit consists of four parts; signal conditioning, tone decoding, display decoding/driving, and the display itself. Let's take each section individually, and discuss the details.

For the input signal, standard audio can be applied via the phone jack J1. No special conditioning is necessary, but the incoming data's potential cannot exceed 5 volts.

As for a telephone interface, the incoming signal has to be regulated by a protective network. As stated earlier, voltages on telephone lines can get quite high, and this can be lethal to delicate ICs. So, varistor V1 is employed to trap spikes and transients, while 1:1 ratio transformer T1 (usually 600 to 600 ohms) is used to isolate the rest of the circuit from the phone line (**Figure 2-2**).

The tone decoder chosen for this project is the 75T202 low power DTMF (Dual Tone Multi-Frequency) receiver. This chip has been around for a while, and always performs reliably. Note that it utilizes a standard color burst crystal (3.579545 MHz), XTAL1, to control the on-board oscillator. Incidentally, the BCD output can address any BCD device. I mention this in the event you want to use it for other purposes.

A 74LS47 BCD to 7 segment decoder/driver does the honors as

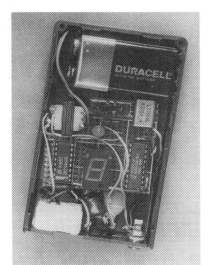

Photo 2-1. The DTMF detector housed in a small project case.

Project #2: DTMF Decoder

Photo 2-2. An external view of the completed DTMF detector.

a display control. It is a simple chip to use, but bear in mind that its outputs go low when activated, thus requiring a common anode configured display unit.

Finally, the display, of the common anode variety, can be any color and just about any size. The 74LS47 has no trouble driving displays as large as 0.8 inches, so this aspect is left up to your needs and/or preference.

As a final "finally", **Figure 2-2** also illustrates the use of a lN4733, 5.1 volt zener diode as a 9 volt battery regulator. This arrangement provides portability, but if you are planning to run the unit off AC wall current, a standard regulated 5 volt power supply is recommended.

The construction method is not at all critical. Breadboard, point-to-point, or PCB wiring can be employed, each providing excellent results. If you go the battery operated route, the unit can be housed in a hand sized case making it very portable, and discrete.

CONCLUSION

Whether motivated by curiosity, pragmatism, or another reason, once completed, this circuit provides the tool to decipher tone pairs. Additionally, it furnishes

Photo 2-3. An example of the piont-to-point wiring method.

some information about digital electronics and associated components. This is an area inescapable these days, but also very productive, as well as fascinating.

Much more will be said throughout this book, regarding digital electronics, and as seen with this project, the resulting devices are well worth the time.

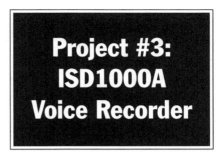

What You Will Need:

✔	U1	ISD1000A Series Voice Record/Play Back IC (Radio Shack, Jameco)
✔	R1	1,000 Ohm, 1/4 Watt Carbon Film Resistor
✔	R2/R4	10,000 Ohm, 1/4 Watt Carbon Film Resistor
✔	R3	470,000 Ohm, 1/4 Watt Carbon Film Resistor
✔	R5	2,2000 Ohm, 1/4 Watt Carbon Film Resistor
✔	C1	1 Microfarad Electrolytic Capacitor (tantalum)
✔	C2/C4	0.1 Microfarad Mylar Capacitor
✔	C3	6.8 Microfarad Electrolytic Capacitor (tantalum)
✔	C5	33 Microfarad Electrolytic Capacitor (aluminum)
✔	SPKR	8 Ohm Speaker
✔	MIC	Electret Microphone
✔	S1	Normally Open Push Button Switch
✔	S2	SPDT Slide or Toggle Switch

Integrated Circuit Projects

Note:

Two sources for digital voice recording chips are given, but others may exist. Check some of the other houses listed in Appendix A and you may find better prices. Even with the ISD chip, this project shouldn't run much over twenty dollars to build.

Forget your pen? Don't having anything to write on? Never fear, the electronic memo pad is here. This device uses one of the most intriguing ICs to come down the pike in a long time. In principle, it is quite simple, but as this discussion will reveal, the versatility and performance are remarkable.

On a single twenty eight pin chip an entire recording system has been placed, requiring just a handful of discrete components to complete a working twenty-second digital sound storage unit. The ingenuity of that system is nothing short of superb, including a preamplifier, filters, automatic gain control, clock, analog storage, control logic, and power amplifier.

If I seem to effuse praise on this chip, it is, at least in my mind, justified. Some time has gone by since I have been this impressed with a single IC. I think once you have put it to use, you too will share my enthusiasm.

CIRCUIT THEORY

While this configuration employs the digital recorder in a basic record/playback mode, the chip possesses the options, to name a few, of memory addressed record and playback, sequential messages, looping, and cascading of additional chips. First, however, lets examine the basic system, then I will discuss the other features.

The secret behind this chip's efficiency is the use of EEPROM cells for storing the analog data (sound). EEPROM stands for Electrically Erasable and Programmable Read Only Memory, which is a type of non-volatile computer memory. Non-volatile refers to the fact that, unlike Random Access Memory (RAM), the memory does not loose its content when power is removed.

Thus, once recorded into the EEPROM, the data will remain for a guaranteed minimum of ten years (probably a lot longer), even if you turn the unit off. Of course, you can record over that data at any time, and the new message will replace the old one.

Additionally, the built-in filters insure optimum sound quality. In fact, digital recordings are far more realistic than analog tape. Nobody ever thinks they sound like themselves, though. This is due, in part, because you hear yourself through the bone and tissue of your head as well as the open air others hear you through.

Anyway, ask anyone. You will sound more realistic when recorded on a digital format than magnetic tape. This is one reason these chips have become a favorite with answering machine designers.

For the sake of a little more detail (you may not want this, but you're going to get it anyway), the chip samples at a rate of 6.4 kilohertz. That is to say that 6,400 bits of data are recorded each second. The total storage area is 128,000 cells, so this allows for twenty seconds of recording time.

Unlike many digital storage systems, the ISD technology eliminates the need for analog-to-digital (A/D), and digital-to-analog (D/A) converters. This not only reduces the complexity and size of the chip, but it also reduces the number of memory cells needed.

In the end, this IC can record the equivalent of one megabit of information on 128,000 bits. This greatly improves the performance of any system using this chip.

Let's take a look at some of the other features incorporated on this IC. One of the most functional utilities is the eight bit address bus (A0 to A7). Through this bus, you can break up the total memory space into segments that can be called on individually. In that fashion, several short messages can be placed, and played back at will.

Some of the pins within this address bus also serve additional functions. For example, if pin A3 is high, when the chip enable (CE) pin goes low, the recorded message will repeat (loop) over and over again. This feature has all kinds of possibilities.

Another pragmatic virtue of this chip is the cascade ability to increase the recording time. Up to sixty seconds can be achieved by using three separate ICs hooked together. All addressing, or other controls, still prevail under this condition.

So, as I promised, this is one handy device. The applications are numerous, and the reliability will be vastly appreciated.

CIRCUIT DESCRIPTION

Since I have called this device "the chip" throughout the first part of this chapter, let me give it a name. It is the ISD1000A Voice Record/Playback IC. Actually, it is one of a series of similar components, all of which accomplish the same goal. Primarily, the difference in each chip is the amount of EEPROM space, which determines the total recording time.

Project #3: ISD1000A Voice Recorder

In addition to the 1000A, this series includes the ISD1420P (20 seconds), the ISD2560 (60 seconds), and the ISD2590 (90 seconds). As seen, each chip progressively increases the record time, with the 2590 providing a full minute and a half.

As for prices, these chips are not exactly cheap. The 1000A and 1420P will set you back between $10 and $15, while the 2560 and 2590 come in at around $35 each. However, you are getting a lot of silicon for the price. Watch for the occasional sales.

Figure 3-1 is the pinout for the ISD1000A, which is packaged in a standard 28 pin plastic DIP (dual in-line pin). Not much

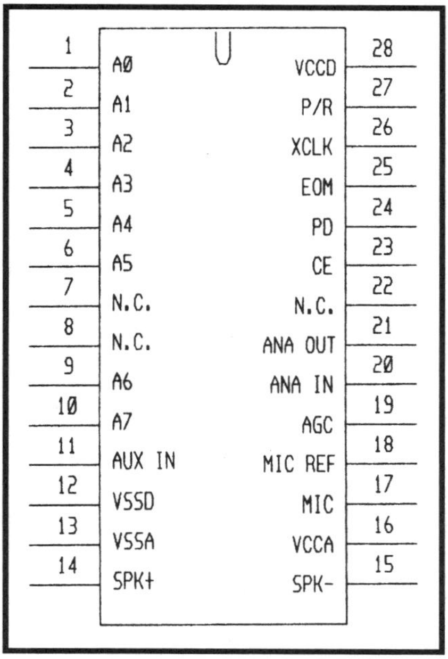

Figure 3-1. ISD1000A voice chip pinout.

```
 1  A0         VCCD  28
 2  A1         P/R   27
 3  A2         XCLK  26
 4  A3         EOM   25
 5  A4         PD    24
 6  A5         CE    23
 7  N.C.       N.C.  22
 8  N.C.       ANA OUT 21
 9  A6         ANA IN  20
10  A7         AGC   19
11  AUX IN     MIC REF 18
12  VSSD       MIC   17
13  VSSA       VCCA  16
14  SPK+       SPK-  15
```

PINS 1-6, 9, 10: Address Bus
PINS 7, 8, 22: Not Connected
PIN 11: Auxiliary Input
PINS 12, 13: Digital/Analog Grounds

PINS 14, 15:	+/- Speaker Connections
PIN 16:	Analog Positive Connection
PIN 17:	Microphone
PIN 18:	Microphone Reference
PIN 19:	Automatic Gain Control
PIN 20:	Analog Input
PIN 21:	Analog Output
PIN 23:	Chip Enable
PIN 24:	Power Down Mode
PIN 25:	End of Message Signal
PIN 26:	Internal Clock Enable
PIN 27:	Play/Record Control
PIN 28:	Digital Positive Connection

Integrated Circuit Projects

needs to be said about this. It's pretty straightforward, except that the chip utilizes a separate analog and digital ground (Vssd/Vssa), and analog and digital positive 5 volt connections (Vccd/Vcca).

The reason for this involves the quality of the audio performance, and if you go with the ISD1000A from Radio Shack, the Archer data book provides excellent information regarding this point. If you purchase the 1420P, 2560, or 2590 from another supplier, like Jameco Electronics, be sure to spend the extra buck for the data sheets. It is well worth the money.

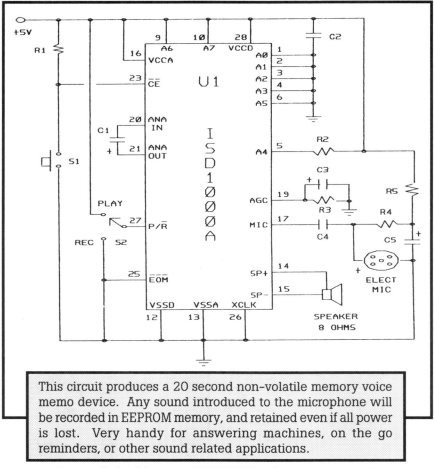

This circuit produces a 20 second non-volatile memory voice memo device. Any sound introduced to the microphone will be recorded in EEPROM memory, and retained even if all power is lost. Very handy for answering machines, on the go reminders, or other sound related applications.

Figure 3-2. *20 second EEPROM voice memo circuit.*

Project #3: ISD1000A Voice Recorder

Analog refers to anything involving the audio sections; microphone input, speaker output, automatic gain control, etc. Digital, on the other hand, concerns addressing and other logic functions. To prevent interference between the two sections keep the components and power leads as separate as possible. Also wire de-coupling capacitor C2 as close to the chip as is practical.

A quick look at **Figure 3-2**, the schematic, reveals the simplicity of this configuration. The few discrete components are standard off the shelf items, and the construction technique can be point-to-point, or PCB. PCB is preferred, however, as it allows for a large ground plane, always advisable in audio circuits.

Maintaining short lead lengths will help prevent audio noise, as well as promote good sound quality. It probably doesn't need to be said, but always use top quality components when constructing your projects.

CONCLUSION

The ISD series provides a hands on preview to a very up-and-coming technology. While digital sound storage is not exactly new (CDs have been around for awhile), the non-volatile memory approach opens up an exciting horizon. The ability to easily change the data in memory has led to an ever increasing use of these chips.

In addition to the pocket voice memo application, initial messages on answering machines, special message greeting

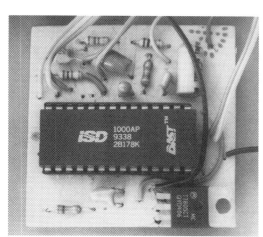
Photo 3-1. The compact circuit board.

Integrated Circuit Projects

cards, announcements on amateur radio systems, voices for childrens' toys, and get your attention in-store advertising are all examples of this technology being put to pragmatic, and often profitable use. And that is just the proverbial tip of the iceberg.

With a little thought, I am confident that you can come up with many other implementations. Think fast, though, as the manufacturing industry is continually employing these chips in a variety of new products, and you don't want them to steal your ideas.

Project #4: Digital Audio Frequency Meter

What You Will Need:

- ✔ U1 — LM386N, Low Voltage Audio Amplifier
- ✔ U2 — LM555CN, Timer/Oscillator
- ✔ U3 — LM7805T, Positive 5 Volt Regulator
- ✔ U4 — LM7905T, Negative 5 Volt Regulator
- ✔ U5 — ICL7107, 3 1/2 Digit LED A/D Converter
- ✔ Q1 — 2N3904 NPN Transistor
- ✔ BR1 — 200 Volt, 1 Amp Bridge Rectifier (W02G, or equiv.)
- ✔ DSPLY — Two Dual Digit Common Anode LED Displays (MAN6710, or Equiv.)

All resistors 5% carbon film.

- ✔ R1 — 1,000 Ohm, 1/4 Watt Resistor
- ✔ R2, 9, 15 — 100,000 Ohm, 1/4 Watt Resistors
- ✔ R3, 6, 7 — 10,000 Ohm, 1/4 Watt Resistors
- ✔ R4, 8, 10 — 10,000 Ohm Potentiometers
- ✔ R5 — 10 Ohm, 1/4 Watt Resistor
- ✔ R11 — 470 Ohm, 1/4 Watt Resistor
- ✔ R12 — 1,000,000 Ohm, 1/4 Watt Resistor
- ✔ R13 — 25,000 Ohm Potentiometer

Integrated Circuit Projects

✔	R14	22,000 Ohm, 1/4 Watt Resistor
✔	R16	470,000 Ohm, 1/4 Watt Resistor
✔	C1	2.2 Microfarad Electrolytic Capacitor
✔	C2	0.05 Microfarad Mylar, or Disk Capacitor
✔	C3	220 Microfarad Electrolytic Capacitor
✔	C4, 5, 12	0.01 Microfarad Mylar Capacitors
✔	C6, 14	0.1 Microfarad Mylar Capacitors
✔	C7, 8, 10	1 Microfarad Electrolytic Capacitors
✔	C9, 11	470 Microfarad Electrolytic Capacitors
✔	C13	100 Picofarad Disk Capacitor
✔	C15	0.047 Microfarad Mylar Capacitor
✔	C16	0.22 Microfarad Mylar Capacitor
✔	MIC	Electret Microphone
✔	S1	SP3T Rotary Switch
✔	MISC	Solder, Hook-up Wire, Knobs, Heat Sinks (Regulators), Etc.

Has there been a time when you needed to know, or were curious about the exact frequency of a certain sound? If the answer is yes, as it has been with me many times, then this project will surely catch your attention. The result will be a 3 1/2 digit device that detects the sound and displays its frequency.

Actually, this circuit can read frequencies much higher than the audio range, but the problem is the microphone. It doesn't detect anything much above 20,000 hertz. So, as designed, we will consider this an audible sound frequency meter.

Project #4: Digital Audio Frequency Meter

CIRCUIT THEORY

In addition to the power supply, this configuration consists of four sections; the pre-amplifier, the audio amplifier, a monostable oscillator, and a digital voltage meter. Each performs a specific function within the total circuit to accomplish the end result.

The pre-amp acts to boost sound detected by the microphone, and condition it for the main audio amplifier (amp). This is a simple one transistor circuit commonly seen in many applications.

The audio amp section further boosts the sound signal. An LM386 low voltage amp is configured to produce a gain of twenty, so the amplifier's output will be about twenty times its input. If even higher gain is required, connect a 10 microfarad capacitor between pins 1 and 8 (+ at pin 1), and this will provide a gain of 200.

Section three utilizes an LM555 timer IC as a monostable oscillator. By routing the audio amp output to the oscillator's trigger, a voltage proportional to the frequency input is generated at the 555's output. The switch provides range control by changing the value of the discharge capacitor, while the potentiometer affords fine tuning.

The last stage furnishes the digital readout via a voltage divider that calibrates the system. The secret here is an A/D converter and LED display driver all wrapped up in one neat little package. This chip, known as the ICL7107, has been around for awhile, but is one of the best digital voltmeters money can buy.

Inside one IC you get the A/D converter, the timing oscillator, digit decoders, and digit drivers. Additionally, the 7107 will provide a true zero reading at zero voltage input. That might not sound important, but it makes calibration a breeze.

One drawback is that like most dual slope D/A converters, the 7107 requires a split +5 volt/-5 volt power supply. That is where ICs U3 and U4 come into play. They provide, in standard linear configuration, both the dual supply, and the +5 volts for the remainder of the circuit.

However, considering all that the 7107 will do for you, this is a small price to pay. The split supply does add a few components to the board, but is not at all complicated.

CIRCUIT DESCRIPTION

A first glance at **Figure 4-1** might lead you to believe this is a complex circuit, but in reality, it's pretty straightforward (all standard applications). Construction options include point-to-point wiring, breadboarding, or PCB, although for a permanent assembly, a PCB will afford the most reliable service.

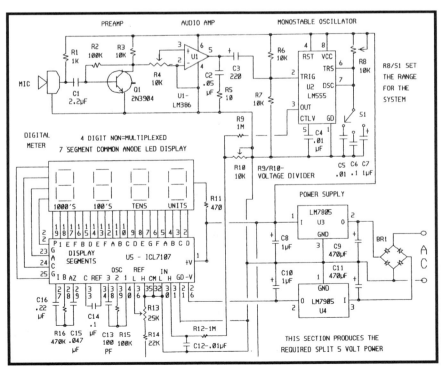

Figure 4-1. LM555 based digital audio frequency meter.

Project #4: Digital Audio Frequency Meter

There is nothing especially critical regarding the layout, except that it is always wise to construct audio projects in a clean fashion, using as large a ground plane (with PCB, leave as much copper as possible where ground connections are made) as is convenient. This will help prevent the circuit from absorbing extraneous noise.

As earlier stated, switch S1 provides three different ranges by changing the value of the discharge capacitor. The larger the capacitor, the lower the frequency the meter will detect. Thus, C7 will be used for 10 hertz to about 6,500 hertz, C6 will cover the mid range of 6,000 to 12,000 hertz, and C5 handles the upper frequencies of 12,000 to 20,000 hertz.

Potentiometer R8 allows for fine tuning of the display, as there is always some tolerance in the value of electronic components. Once the meter is calibrated, this control can be used to individually adjust the accuracy of each range.

As for calibration, it will require a sound standard of known frequency. A LM555 astable oscillator is one such source, or an audio frequency standard, say 1 kilohertz, would be another. If you have access to a bench frequency meter or oscilloscope, this will be valuable in confirming the output of the audio standard.

However, one initial adjustment needs to be made before we calibrate the sound section of the circuit. This is the reference voltage of the digital meter, or more specifically, potentiometer R13. The 7107 wants to see one half the full scale voltage as the reference voltage. In our case, full scale voltage is 2 volts, thus the reference voltage needs to be 1 volt.

This is controlled by R13, and can be measured at the high and low reference pins (35 and 36), but a easier way is to introduce a voltage of known value, 1.5 volts for example, to the 7107 high and low inputs (pins 30 and 31), then adjust R13 to match that voltage. This second method is faster and just as accurate, provided you are certain about the input voltage value.

With the voltage reference set, we can move on to calibration. Connect the frequency standard's output to a speaker, set the meter's range, and allow the microphone to pick up the sound. First use potentiometer R10 to get the display reading in the ball park, then fine tune the reading with R8.

Repeating this process for several different frequencies and range settings will fully calibrate the device. With this done, the audio frequency meter is ready to take around measuring sounds. It will come in handy whenever you need to know that information.

CONCLUSION

The digital audio frequency meter furnishes insight into several areas of electronics. Here are mini-lessons in audio amplification, monostable multivibrators (oscillators), and A/D conversion. It also will arm you with knowledge that will prove to be very useful.

Additionally, each stage of this circuit can be employed in projects of a completely different color. A little modification can give birth to a multitude of pragmatic, interesting and educational endeavors. Only your imagination limits you.

As for the sound meter itself, it will serve you well. Some applications that immediately come to mind are reading telephone DTMF tone pair (the unit will average the two tones), confirming audio encoder tones, tuning a guitar or other string musical instrument, testing audio frequency generators, or checking anything that makes a specific noise.

Project #4: Digital Audio Frequency Meter

As long as you know what frequency the sound is supposed to be, you can easily verify the accuracy of said sound. This can be an important tool in troubleshooting equipment controlled by audio tones. By the way, employing an ultrasonic transducer, in place of the microphone, will extend the meter's capability into that range.

So give this one a try. I don't think you will be disappointed. But even if you are, think of all the good stuff you have learned along the way.

Integrated Circuit Projects

Project #5: Temperature Detector

What You Will Need:

- ✔ U1 — LM741CN, Operational Amplifier
- ✔ LED1 — T-1 3/4 Jumbo Light Emitting Diode (optional)
- ✔ R1 — 10,000 Ohm, 1/4 Watt Resistor
- ✔ R2 — 100,000 Ohm Potentiometer
- ✔ THM1 — Bead Style Negative Coefficient Thermistor (All Elect.)
- ✔ RLED — 470 Ohm, 1/4 Watt Resistor
- ✔ RLY — Relay (voltage will depend on operation voltage)
- ✔ PIZO — Any Piezo Buzzer (driver built in)
- ✔ LED — Standard LED/Dropping Resistor Series Network

This is an extremely simple circuit that utilizes almost any operational amplifier (op-amp) to monitor changes in temperature, and it is also one of my favorites. So, I had to include it. I have literally lost count of the number of times I've employed this circuit; it's that useful.

All that is required is the op-amp, two resistors, and a thermistor. As stated, virtually any op-amp will do, but the LM741 performs well, and it's cheap. Speaking of which, the entire monitor can be constructed for around two dollars.

Once completed, this unit will very accurately keep an eye on whatever temperature you set, and alert you to a rise above that setting. By simply reversing the inverting and non-inverting inputs of the op-amp, the device will then alert you to a drop below that temperature.

I think you can already visualize the value of this circuit. In any situation where a change in ambient temperature is important, this is your boy.

CIRCUIT THEORY

In the configuration, illustrated in **Figure 5-1**, an LM741 (op-amp) is set up as a comparator. More specifically, a very tightly controlled window comparator. The principle here is to establish a range, or window, in which the op-amp output remains low by governing the voltage at both the inverting and non-inverting inputs.

The non-inverting voltage is set by potentiometer R2, while the inverting potential is the result of the voltage divider R1 and thermistor THM1. With R2 adjusted, as long as this equilibrium is maintained, the op-amp output remains low (off). However, if either voltage changes, the window is exceeded, and the op-amp output goes high (on).

Project #5: Temperature Detector

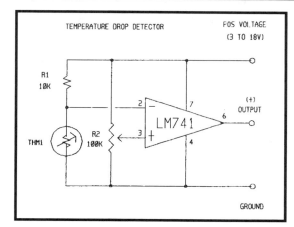
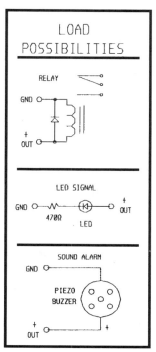

This simple circuit can be used to detect a change in the ambient temperature around the probe. The probe is a negative coefficient bead style thermistor encapsuled in epoxy, or some other protective material. As seen here, the circuit will detect a rise in temperature above a set-point, but by reversing leads 2 and 3 of LM741, a drop in temperature can be detected. Virtually any standard op-amp can be employed, and the loads shown at right are a few suggested devices.

Figure 5-1. LM741 based temperature detector.

Of course, what makes this all work is that the thermistor's resistance will change in response to a temperature change. When that happens, the equilibrium is disrupted, and the op-amp output goes high. That, in turn, sets off whatever indicator or alarm you have connected to the output (**Figure 5-1** suggests a few).

CIRCUIT DESCRIPTION

Considering the simplicity, you ought to be able to go into business and turn out eight or ten of these a day. That's if you are so inclined. The circuit can be easily constructed on perf-board or PCB material, and can be kept quite small in size.

Integrated Circuit Projects

As I said earlier, **Figure 5-1** illustrates the device as a rise-in temperature monitor, but it is a simple matter to reverse the op-amp inputs (pins 2 and 3) and detect drops in temperature. Either way, you will need to expose the thermistor to the temperature you want to monitor, then adjust R2 so that the op-amp just turns off.

Now, when the temperature rises or falls below the setting, the output will activate an LED relay, buzzer, or whatever device you have connected. This is good! It has many possibilities.

One tip; if you plan on putting the sensor (thermistor) outside, or in liquids, be sure to protect it. I usually encapsule it in epoxy cement, or fiberglass resin. If it is going to monitor high temperatures, encase it in metal tubing or something that can stand the heat.

The operating voltage can be from 3 volts to 18 volts, at least with the LM741, so this leaves excellent latitude if you are adding this circuit to existing equipment. Current drain is very low, probably no more than 10 milliamps. Again, this depends on the op-amp used. With an LM358AN, for example, that drain is going to be considerably less.

CONCLUSION

So, there it is. The super simple op-amp temperature monitor. As I said, it is one of my favorites. This project first came about as the result of a broken water pipe due to freezing weather. Brother, what a mess. I mean water every ... well, you don't even want to hear about it. I decided that was not going to happen again, so I rigged up this circuit to let me know when the temperature reached freezing.

Project #5: Temperature Detector

It worked. Haven't had a broken water pipe since, but that, of course, is not the only use for this gem. It can be employed in regulating any vessel of liquid, say a water bath for photographic purposes, or the tank temperature of your prized Samurai Double Lipped Fighting Fish.

In the temperature rising mode, it could save you a small fortune in frozen food products by alerting you to a problem with your freezer. Or, how about letting you know that you are about to burn up your computer's expensive Pentium 166 processor? The monitor could also warn of dangerously high temperatures in a water heater or furnace system. Just be sure you properly protect the thermistor at higher temperatures. The epoxy, for instance, will begin to melt at about 300 degrees Fahrenheit.

These are just a few of the possible applications. With a little thought and/or need, I know you can come up with others.

Integrated Circuit Projects

Project #6: TTL Counter

What You Will Need:

- ✔ U1 — LM555, Timer/Oscillator
- ✔ U2 — 74LS90, Decade Counter
- ✔ U3 — 74LS47, BCD to 7 Segment Decoder/Driver
- ✔ DSPL1 — Common Anode, 7 Segment LED Display

All resistors 5% carbon film.

- ✔ R1 — 1,000,000 Ohm Potentiometer
- ✔ R2 — 1,000 Ohm, 1/4 Watt Resistor
- ✔ R3 thru 9 — 470 Ohm, 1/4 or 1/8 Watt Resistors
- ✔ C1 — 220 Microfarad Electrolytic Capacitor
- ✔ C2 — 3.3 Microfarad Electrolytic Capacitor
- ✔ C3 — 0.1 Microfarad Mylar Capacitor

Integrated Circuit Projects

The realm of digital electronics has many avenues, but one of the most interesting, and ultimately practical is that of counters. The ability of an electronic circuit to keep track of pulses is a fascinating area of experimentation, not to mention an industrial fundamental.

The computer world would be hopelessly adrift in a quagmire of confusion if it were not for devices that either create pulses or count them. For that matter, without them, there would not be a computer world, as television, radio, Fax machines, telephones, digital clocks, and a host of other commonly admired creature comforts, all are, to some degree, dependent on this technology.

So, for this reason alone it is advisable to have a least a basic understanding of counters, and there isn't a better place to start than with this project. A bonus is that it's fun to build.

CIRCUIT THEORY

This circuit is designed with an LED display to provide a visual method of following its operation. Naturally, not all counting circuits have, or need this feature, but for our purpose, as with many applications, the display is necessary.

There are three sections to this project, the timer (clock), the counter/binary encoder and the binary decoder/display driver. As seen in **Figure 6-1**, the circuit has a linear configuration with the LM555 timer driving the 74LS90 counter which, in turn, fuels the 74LS47 driver and display. Let's take each section individually.

The timer/clock is a standard LM555 astable configuration with Rl and Cl/C2 determining the frequency output. For this circuit, two discharge capacitors (Cl/C2) are employed providing both

Project #6: TTL Counter

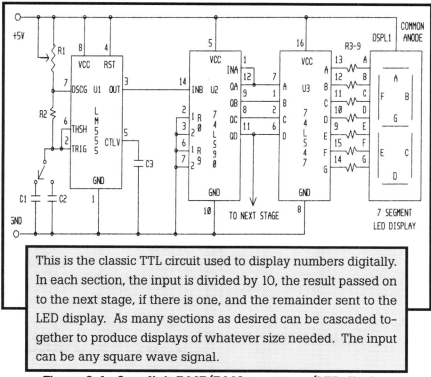

This is the classic TTL circuit used to display numbers digitally. In each section, the input is divided by 10, the result passed on to the next stage, if there is one, and the remainder sent to the LED display. As many sections as desired can be cascaded together to produce displays of whatever size needed. The input can be any square wave signal.

Figure 6-1. One digit 7447/7449 counter w/LED display.

second and minute counts. The larger capacitor C1 furnishes the minutes, while seconds are controlled by C2. Potentiometer R1 acts as the fine tuning, and will adjust the timer to the exact pulse.

The 74LS90, officially know as a Decade Counter, actually performs two functions. First, the chip counts the pulses from the 555 timer, then it encodes those pulses into a binary 4 bit output. Each decimal digit has its very own binary code consisting of zeros and ones **(Table 6-1** illustrates 0 to 9), and it is this code that tells the 74LS47 (the next section) what to do.

For example, the binary representation of the decimal digit "5" is "0101" which means that the 74LS90 outputs will be high (1), low (0), high and low respectively (note that binary numbers read backwards. Cute, isn't it?). Anyway, that 4 bit array means "5" to any BCD device.

Integrated Circuit Projects

DECIMAL	BINARY
0	0000
1	0001
2	0010
3	0011
4	0100
5	0101
6	0110
7	0111
8	1000
9	1001

Table 6-1. Binary number codes for decimals 0 to 9.

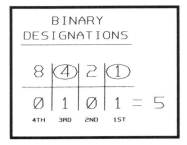

BINARY DESIGNATIONS

8 ④ 2 ①
0 | 1 | 0 | 1 = 5
4TH 3RD 2ND 1ST

As seen in the above example, the four digit binary representation of the number 5 is 0101. Remember that the binary numbers read from right to left, and the total is the sum of all individual digits represented by a "1". For "5", the third, or "4" digit, is added to the first, or "1" digit, bringing the total to five.

The 74LS47 is a BCD device, so when it sees "0101" at its inputs, it knows to instruct the display, through the on-board display drivers, to produce the number five. This is accomplished by illuminating the appropriate LED display segments.

If you look closely at **Figure 6-1**, you will see that the LED display is comprised of seven individual sections, or segments, and that they are designated A, B, C, D, E, F and G (center). With these seven segments arranged as they are, any digit can be represented. For a "5", the 74LS47 activates the A, C, D, F, and G sections. If a "2" is needed, A, B, D, E and G would be energized, while an "8" would require all seven segments.

The last components that bear mentioning are resistors R3 to R9. These are dropping resistors that protect the LED display against excessive current and voltage. Levels too high will damage, or burn out the segments; thus these are essential.

Finally, the TTL series of ICs has a very narrow voltage tolerance (4.75 to 5.25 volts), so a regulated five volt power supply is highly recommended. Not only will this keep the supply power in line, but it will also keep the, performance of the circuit stable and accurate.

CIRCUIT DESCRIPTION

As with most projects in this book, the counter can be constructed with point-to-point wiring, PCB or breadboard technique. The configuration is simple, and not at all critical but, as always, work carefully.

As stated earlier, the speed of the 555 clock is determined by discharge capacitors C1 or C2, and potentiometer R1. The larger the capacitor, the longer the pulse will be; thus, a great deal of lattice exists in this area. once the general interval is selected by the capacitor value (minutes, seconds, etc.), the exact time period can be set with R1.

Referring to **Figure 6-1**, you'll notice a vertical arrow designated "To Next Stage". You will also notice that the arrow taps off from the 74LS90 pin 11, to 74LS47 pin 6 connection. It is from this point that you can cascade additional counter stages into the circuit.

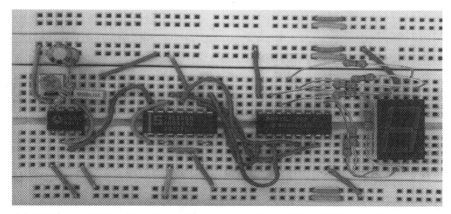

Photo 6-1. Solderless breadboard version of the TTL counter circuit.

As many new digits as desired can be added by merely running the pin 11/6 line to pin 14 of the next 74LS47. The clock (555) only needs to be connected to the first counter, and the rest of the circuit remains unchanged.

Since a single digit is somewhat limited, this is a handy feature. Each new stage will represent ten time the previous, thus a four digit system, set for seconds, would have a full scale display of "9999" seconds, or just over 2 3/4 hours counting time. In the minutes setting, that would increase to 166.65 hours, or just short of a week.

To further expand on this, changing the discharge capacitor to 2,200 microfarads will produce "ten minute" counts. In that scenario, a "9999" reading would represent seventy days, or a little less than one fifth of a year.

Regarding layout; since the chip designers were nice enough to place all the 74LS47's outputs on one side of the IC, interfacing to the display is a simple matter. The dropping resistors can be as small as 1/8 watt, if space is a consideration, but even with 1/4 watt resistors, the completed circuit won't take up a lot of space.

CONCLUSION

Before you lay the secrets of the digital counter. This highly functional device can be of immense assistance to existing equipment, and prototypes alike. Any project that requires a visual display of its clock pulse can make noble use of this circuit, as the timing signal can be replaced by any source producing square wave, or even sawtooth signals.

If you are looking for extreme accuracy, a crystal controlled clock can be substituted for the LM555 timer. In that fashion, the pulses are controlled by a quartz crystal, usually to a tolerance of five hundreths of a percent. However, I think you will discover that for most applications, the 555 does a more than admirable job.

On a last note, one of the most commendable aspects of this design is its flexibility. I have mentioned several ways you can stray from theoriginal configuration and end up with a specialized version of this device, but I am confident you will invision additional modifications. So, don't disappoint me! Put this one to good use.

Integrated Circuit Projects

Project #7: Up/Down Counter

What You Will Need:

✔	U1	LM555, Timer/Oscillator
✔	U2	74LS193, Binary Up/Down Counter with Clear
✔	U3	74LS47, BCD to 7 Segment Decoder/Driver
✔	DSP1	Common Anode, 7 Segment LED Display

All resistors 5% carbon film.

✔	R1	1,000,000 Ohm Potentiometer
✔	R2	1,000 Ohm, 1/4 Watt Resistor
✔	R3 thru 9	470 Ohm, 1/4 or 1/8 Watt Resistors
✔	C1	220 Microfarad Electrolytic Capacitor
✔	C2	3.3 Microfarad Electrolytic Capacitor
✔	C3	0.1 Microfarad Mylar Capacitor
✔	S1	SPDT Toggle, Slide, or Other Type Switch
✔	S2	DPDT Toggle, Slide, Etc.

Integrated Circuit Projects

In chapter six, we built the classic single digit counter using the 74LS90 and 74LS47 chips. This circuit has become, over the years, an Industry Standard, but it does have one drawback. It will only count up, resetting to zero on the pulse following the full scale count of "9".

But, what if you would like to count up, then back down, or change the counting direction at any given point? Well, guess what, the TTL series has an answer for that, too. It is the 74LS193, Binary UP/Down Counter, and is just as easy to use as the 74LS90.

The major difference, as far as we are concerned, is that 74LS193 has both an UP, and a DOWN input. By throwing the clock to one, and positive 5 volts to the other, you can force the count to either rise or fall.

Actually, appropriate application of the +5 volts is what does the trick, but more about that in a moment. If you are looking for the ability to control the counter's direction, the 74LS193 is the ticket. Let's take a look at how we can plug it into the previous counter circuit.

CIRCUIT THEORY

A quick glance at **Figure 7-1** reveals a circuit almost identical to the TTL counter in the last chapter. The primary difference is the double pole-double throw switch (DPDT-S2), between the clock and the up/down counter. This switch is used to control the direction of the count by controlling the destination of the clock signal and positive five volts.

The way the two 74LS193 inputs function requires that one receive the clock pulses, while the other is tied to the positive rail. Which ever is high (+5 volt) determines the count direc-

Project #7: Up/Down Counter

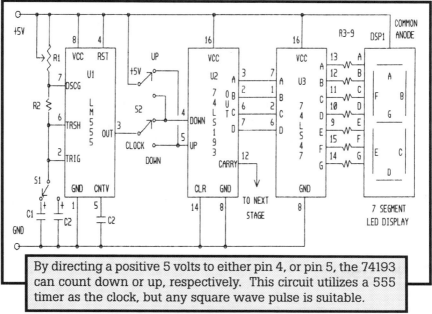

By directing a positive 5 volts to either pin 4, or pin 5, the 74193 can count down or up, respectively. This circuit utilizes a 555 timer as the clock, but any square wave pulse is suitable.

Figure 7-1. The 74193 up/down counter.

tion. So, when pin 5 is high, the count is up, and with pin 4 high, the count is down. In either case, the other input (pin 4 or pin 5) is used to introduce the clock signal.

Beyond that, the 74LS193 functions just like its cousin the 74LS90. Well, I take that back. There is one other difference, and that is pin 12. With the 74LS193, pin 12 is the "carry" line, and this is used to cascade additional chips.

If you remember from chapter six, the 74LS90 simply taps off from its D output (pin 11) for cascading, so this is a unique approach (not terribly unique, however, as "carry" lines are common among many counter chips).

Anyway, beyond that, the circuits perform identically. The next stage is the 74LS47, and we all know what it does. We don't?! OK, I'll refresh your memory.

The 74LS47 decodes the counter's BCD output, and tells the display which segments to turn on. Now, you knew that, didn't you?

In both cases, a common anode display is employed, and R3 through R9 act as dropping resistors. Last, but not least, a regulated five volt power supply is required to meet the TTL requirements and maintain circuit stability.

So, if you haven't yet torn the last circuit apart, you are more than halfway there. Swap out the counter chips, make a few wiring changes, and BINGO, you have an Up/Down counter.

CONCLUSION

Like the 74LS90 circuit, this one can very useful in situations where a counter is needed. Of course, it can function as just an UP counter, if that is all you require, but its ability to count backwards will definitely come in handy with certain applications.

All cascading rules applicable to the 74LS90 apply to the 193. The carry line goes to the appropriate input of the next stage (UP or DOWN), and it is necessary to have all 74LS193s counting in the same direction. Trust me, things can get quit interesting if one chip is counting up, while another is counting down.

There you have it. A digital UP/DOWN counter for your arsenal of pragmatic ICs. I was once browsing through a library book on advanced transistor circuits (circa early 1960's), when I came across a similar circuit. This was described as a compact UP/DOWN counter that could be built on a five by eight inch PCB.

The configuration utilized thirty-nine transistors, seven diodes, and literally dozens of capacitors and resistors. It was state-of-

Project #7: Up/Down Counter

the-art for its time, but just seven years later, that entire circuit could be bought in a package roughly 1 inch by 3/8 of an inch, by 3/16 of an inch, and for less than a dollar.

This, of course, is due to the invention and perfection of the IC by Bell Laboratories. Imagine what Alexander Bell, or Nikola Tesla, or even Albert Einstein would have thought of this technology. The capability, using lasers, microphotography, and other remarkable techniques, of putting tens, hundreds, thousands, even millions of semiconductors on a tiny piece of silicon.

What You Will Need:

✔	U1	LM7805T, Positive 5 Volt Regulator
✔	U2	LM555, Timer/Oscillator
✔	U3	CD4069, Hex Inverter
✔	U4	CD4011, 2-Input Quad NAND Gate
✔	U5	ICM7217BIJI, 4 Digit Up/Down Counter (Mouser, Jameco)
✔	DSPLY	4 Digit, Common Anode Multiplexed LED Display (see notes)

All resistors 5% carbon film.

✔	R1	1,000,000 Ohm Potentiometer
✔	R2	470,000 Ohm, 1/4 Watt Resistor
✔	R3	1,000 Ohm, 1/4 Watt Resistor
✔	R4	10,000 Ohm, 1/4 Watt Resistor
✔	R5	470 Ohm, 1/4 Watt Resistor
✔	C1	1 Microfarad Electrolytic Capacitor
✔	C2	0.01 Microfarad for 1/100 of a Second (1 minute scale)
		1 Microfarad for 1 Second (1 hour scale)
✔	C3	0.01 Microfarad Mylar Capacitor

Integrated Circuit Projects

✔ S1   SPST Toggle, Slide, Other Switch
✔ S2   SPST, Normally Open, Push Button Switch

Note:

The term multiplexing, when applied to an LED display, refers to all like segments (A's, B's, C's, etc.) being tied together. The resulting display has one line for A segments, one line for B segments, and so forth, and a line for each common element (cathode, or anode) of each display. These can be found on the surplus market, in obsolete equipment, or can be made from individual displays. Some good starting places to look for surplus multiplexed displays are the Electronic Goldmine, Alltronics, All Electronics, and B. G. Micro, Inc. (see Appendix A). Unfortunately, Liquid Crystal Displays (LCD) have somewhat replaced LED displays, so the multi-digit variety are getting harder to find.

The ICM7217 series of Up/Down, CMOS Counter/Display Drivers is made by Harris-Intersil, and provides the heart of a highly versatile 4 digit counting system. Combined with four other chips, and a handful of discrete components, this IC becomes a fully functional digital stopwatch.

There are four varieties of the 7217, designated 7217IJI, AIPI, BIJI, and CIPI, and each has its own characteristics. For starters, the IJI and AIPI provide decade counts of "9999" full scale, while the BIJI and CIPI result is timer displays of "5959" full scale.

Additionally, IJI, and BIJI devices drive common anode displays, and the AIPI, and CIPI chips activate common cathode LED indicators. Thus, circuit flexibility is more than obvious.

Project #8: ICM7217 Digital Stopwatch

In this chapter, we will utilize the 7217BIJI to construct a four digit timing unit that accurately measures "59.59" seconds (1 minute), or "59.59" minutes (1 hour). In either case, the count can be reversed, or reset, and those ranges lend themselves well to timing anything from your three minute egg to sporting events.

CIRCUIT THEORY

Like many of the Intersil chips, this one combines a whole lot of stuff in a single package. With the 7217 you get the four decade presettable Up/Down counter, on-board multiplexing oscillator, counter registers, and drivers for a multiplexed LED display (these will handle display sizes up to .8 inches).

Other nice features include a Schmitt trigger on the count input for greater reliability, TTL compatible BCD I/O port, static discharge protection, and continuous comparison of registers to counter. While all this may not sound terribly important, it can be useful in modification, and when handling the chips, helps safeguard the delicate CMOS architecture from dangerous static.

As for the stopwatch itself, this is how it works. Like the previous counter circuits, a clock pulse is furnished by a standard LM555 astable oscillator circuit, but this signal has to control more than just the count input (pin 8), see **Figure 8-1**. From the clock, the configuration must provide the proper signals for the store input (pin 9), and reset (pin 14), and this is accomplished by two logic chips, U3 and U4.

U3 is a CD4069 Hex Inverter, while U4 is a CD4011 Quad NAND Gate, and between the two, the appropriate signals are generated. For example, the store pin wants to see a low (0) in order to transfer the count to the register. Refer to **Figure 8-1** again and you will see that the 555 output (a low going pulse) is

Integrated Circuit Projects

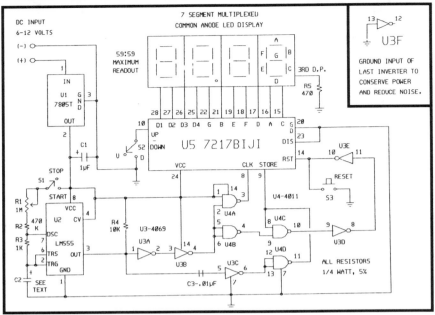

Figure 8-1. The ICM7217 based digital stopwatch.

sent, among other places, to U3c, which inverts it to a high. That high is applied to U4d, a NAND gate wired as an inverter (both inputs tied together), changed to a low, and sent to the 7217's store pin.

What this means is that each time a new timer pulse is detected, the logic chips deliver a low to U5's pin 9, which transfers the last count to the registers. The registers, in turn, instruct the drivers to display that count. Consequently, the display is updated with each new pulse.

The reset (pin 14) also requires a low to reset the counters, and the count input (pin 8) increments the count on the rising edge (high) of the signal. If you work through the U3/U4 logic route for reset and count input, as we did with store, you will see that at the appropriate times, both receive the desired signals.

This whole logic scheme may seem somewhat complex, but in reality, it is the simplest way to create the necessary highs and

Project #8: ICM7217 Digital Stopwatch

lows at the necessary times. Additionally, with everything connected as per **Figure 8-1**, only one inverter gate, U3f, is left unused, and this supports a discipline everyone should adopt regarding the use of logic related ICs. That is, always try to leave as few unused gates on the board as possible. I hate to waste silicon!

With all the proper signals present, the 7217 drivers activate the LED display. This chip uses what is termed a multiplexed display, which means that all A segments, all B segments, all C segments, etc. are tied together and connected to a single pin.

Each of the four digits has a common anode, connected to its own pin, thus the chip/display interface consists of eleven lines. This arrangement allows for a far more convenient method of addressing the display, since to individually acknowledge all the segments and anodes of a four digit display would require thirty-two lines.

So, you may be asking yourself, "How does this work?". Well, it isn't all that mysterious. Since each of the segments is tied to a common line, the anodes determine where the segments will light (in which digit). For example, if the A segment needs to be activated in the third digit, the 7217 will engage pins 16 and 26. This will provide a conductive path through the third digit's A segment, thus illuminating it.

Subsequently, if the A segment is needed in both the third digit and first digit, the chip will activate pins 16, 26 and 28. That action will light the A's in digit one and three. In this fashion, the 7217 can control all segments in each of the four digits, and do so with a minimum of interconnections.

Power for this circuit is regulated by U1, an LM7805T 5 volt positive regulator. Unlike the more common logic CMOS chips, the 7217 requires a voltage between 4.5 volts and 5.5 volts (the CD4000 series can handle from 3 to 18 volts). For this reason, and for circuit accuracy and stability, the regulated supply is utilized.

Integrated Circuit Projects

That about wraps up circuit theory. The stopwatch uses five ICs, but that doesn't make it complicated. Of course, having all those great features and functions jammed into one 28 pin chip sure helps. See **Figure 8-2** for pin designations.

CIRCUIT DESCRIPTION

Layout is not critical, so either point-to-point wiring, or PCB technique can be used. However, for a permanent installation, PCB is recommended, as it will provide a more durable, thus dependable, finished product.

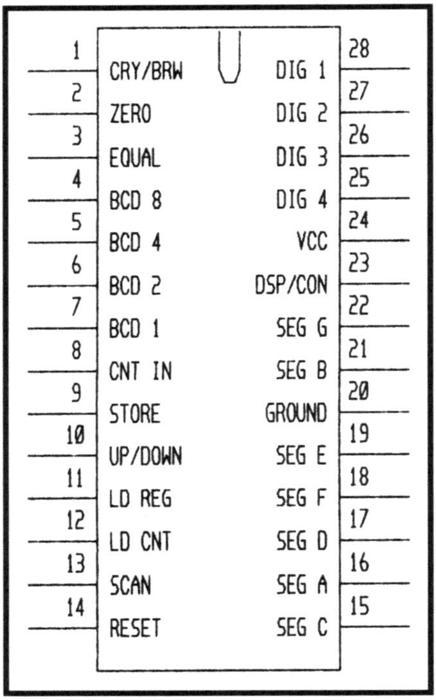

```
 1  ─ CRY/BRW    DIG 1 ─ 28
 2  ─ ZERO       DIG 2 ─ 27
 3  ─ EQUAL      DIG 3 ─ 26
 4  ─ BCD 8      DIG 4 ─ 25
 5  ─ BCD 4      VCC   ─ 24
 6  ─ BCD 2      DSP/CON ─ 23
 7  ─ BCD 1      SEG G ─ 22
 8  ─ CNT IN     SEG B ─ 21
 9  ─ STORE      GROUND ─ 20
10  ─ UP/DOWN    SEG E ─ 19
11  ─ LD REG     SEG F ─ 18
12  ─ LD CNT     SEG D ─ 17
13  ─ SCAN       SEG A ─ 16
14  ─ RESET      SEG C ─ 15
```

PIN 1:	Carry/Borrow Output
PIN 2:	Zero Output
PIN 3:	Equal Output
PIN 4:	BCD Input/Output Eights
PIN 5:	BCD Input/Output Fours
PIN 6:	BCD Input/Output Twos
PIN 7:	BCD Input/Output Ones
PIN 8:	Count Input
PIN 9:	Store Input
PIN 10:	Up/Down Control
PIN 11:	Load Register/Off
PIN 12:	Load Counter/ I/O Off
PIN 13:	Scan Control
PIN 14:	Reset Control
PIN 15:	Display Segment C
PIN 16:	Display Segment A
PIN 17:	Display Segment D
PIN 18:	Display Segment F
PIN 19:	Display Segment E
PIN 20:	Ground
PIN 21:	Display Segment B
PIN 22:	Display Segment G
PIN 23:	Display Control
PIN 24:	Positive 5 Volts
PIN 25:	Digit Driver 4
PIN 26:	Digit Driver 3
PIN 27:	Digit Driver 2
PIN 28:	Digit Driver 1

Figure 8-2. Intersil/Harris ICM72117BIJI pinout diagram.

Project #8: ICM7217 Digital Stopwatch

Construction is straightforward, and common sense should prevail in reference to parts positioning, wiring and other aspects of production. It goes without saying that clean layouts are conducive to reliable operation.

Looking at **Figure 8-1**, you will notice U3f boxed off, and that its input (pin 13) is tied to ground. This is another advisable practice when dealing with logic chips, and that is to connect all unused inputs to either ground or the positive rail. Doing so will not only conserve power, but also prevent the chip from picking up unwanted noise.

Once the circuit is wired and functional, you have three operational controls governing the counter's performance. Theses are the count direction, S2, starting and stopping the count, S1, and resetting the count, S3.

Pin 10 directs the count either UP, if tied to the positive voltage or left floating, or DOWN, if sent to ground. The direction can be changed at any time during the counting operation.

Pin 14, when taken low, resets the counters to zero. Again, this can be done at any time, but with the one second rate, there will be a slight delay, as the reset doesn't take effect until the next pulse.

The stop/start function merely involves disconnecting the clock from the circuit. Switch S1 does just that, and when open, the display will freeze at the last count.

CONCLUSION

Once assembled, this handy-dandy UP/DOWN counter will prove very useful for a variety of purposes. As earlier stated, it can be employed for track and field events, timing recipes, photographic work, or any other situation that requires accurate measurement of elapsed time.

Integrated Circuit Projects

Additionally, the BCD inputs can be used to preset a time period, and then count down from that setting. This is usually done with BCD encoded thumb switches, but it could also be computer controlled.

By replacing the LM555 clock with a gated, or window timing circuit, the unit will function as a frequency counter. The 7217 is guaranteed to 2 megahertz, but usually will reach 5 megahertz without any problem. A simple frequency divider, fashioned from a CD4017 (see chapter 10), can extend that response into the 20 megahertz range.

So, the options for modification are plentiful. And these are just two suggestions. With some thought, I'm confident you can outdo me. Go ahead, give it a shot.

What You Will Need:

- ✔ U1, 2 — 75491, Quad Segment Drivers
- ✔ U3 — 75492, Hex Digit Driver
- ✔ U4 — MK50250N, 4/6 Digit Alarm Clock (Jameco)
- ✔ Q1 — 2N3904 NPN Transistor
- ✔ DSPLY — 4 or 6 Digit LED Clock Style (w/colon) Display
- ✔ D1 — 1N4001 Rectifier Diode
- **All resistors 5% carbon film.**
- ✔ R1 thru 8 — 220 Ohm, 1/4 Watt Resistors
- ✔ R9 — 470 Ohm, 1/4 Watt Resistor
- ✔ R10 — 100 Ohm, 1/4 Watt Resistor
- ✔ C1 — 0.005 Microfarad Disk or Mylar Capacitor
- ✔ S1, 6 — SP3T Rotary or Slide Switches
- ✔ S2, 4, 5 — Momemtary Contact, Normally Open Push Button Switches
- ✔ S3 — SPST Toggle, Slide, Push Button Switch
- ✔ SPKR — Small 8 Ohm Speaker

Integrated Circuit Projects

Notes:

The power supply can be any single positive voltage between 9 and 18 volts. However, a 12 volt system will probably be easiest to provide, and is a good middle of the road value. It can be regulated if desired, but this is not required. Basically, a transformer, bridge rectifier and 470 Microfarad filter capacitor will do quite nicely.

To simplify this project, the Mostek Corporation comes to the rescue. This company markets a wide variety of special purpose ICs, the MK50250N among them. Here we have a complete clock circuit on one chip that offers simple setting, 4 or 6 digit capability, intensity control and a variety of alarm modes, all off a single positive 9 to 18 volt DC power source.

The addition of three ICs, a speaker, a 4 or 6 digit multiplexed display, a few switches and the power supply turns the MK50250 into a full function digital alarm clock. While much of the circuitry is, contained in the MK50250, this still is an instructive and challenging project.

CIRCUIT THEORY

The principle behind any time measuring device, or clock, is to either mechanically or electronically count seconds. I know, some clocks have 1/10 and even 1/100 of a second indicators, but these are bells and whistles. The real task is to keep track of passing seconds, then progressively convert them to minutes and hours.

Conventional (non-electronic) clocks accomplish this with motors, or springs, gears, and other neat mechanical stuff, and perform their task well. The accuracy with all but the cheapest devices is excellent, but as with analog versus digital meters,

Project #9: Digital Alarm Clock

reading a dial can be somewhat more inconvenient, and perhaps less accurate, than reading a clearly indicated numerical display (10:45 for example).

While electronic time keepers have by no means replaced conventional instruments, their popularity is well established. Since their introduction in the late 1960's, digital clocks, wrist watches, kitchen timers and such, abound everywhere you look (well, almost everywhere). This is due as much to their functionality as price or accuracy.

As for operation, it is essentially the same as their mechanical counterparts; keeping track of the seconds. However, unlike their mechanical cousins, this is done through electronics, not gears, much of which is common to many other areas of electronics.

In almost every case, the starting point is a crystal controlled oscillator whose fundamental frequency is divided down to one hertz. Most wrist watches utilize a 32.768 kilohertz quartz crystal, also know as a tuning fork crystal. This establishes the initial frequency which can now be divided to the needed pulses.

Another popular approach is the MM5369AA/N Programmable Oscillator/Divider (60 Hz). This handy IC uses the readily available 3.579545 megahertz color burst crystal, and generates a 60 hertz output (it also produces a 3.579545 MHz output, but that is primarily for fine tuning purposes).

The 60 hertz is then sent to a divide-by-six divider for a 10 hertz output which is, in turn, taken down to a 1 hertz pulse by a divide-by-ten stage. Thus, this enterprising chip provides a stable, reliable and inexpensive source of a one second pulse.

A third method will only work with clocks that are powered by standard 120 volt AC lines, but it is quite clever. Instead of generating a sixty hertz signal with an MM5369, or other configuration, why not take it off the AC line? After all, household current is sixty hertz, isn't it? Why not, indeed. By tapping into the secondary side of the power transformer, the initial sixty hertz is obtained, and can be divided down from there. See, I told you it was clever.

OK, with said pulse at hand, some specialized counters now come into play. The first has to count to sixty, trigger the next counter, and reset. Counter two is also a sixty count device which again triggers the third counter, and resets.

The last counter will progress to twelve, or twenty four before resetting. The total count depends on whether the clock is set for standard or military time. In either case, this usually is the last step in the counting process.

Additionally, the second sixty count device has to know to reset the display on the sixth count, as this is the minutes indicator. The last time I looked there was no such time as 6:75, or 10:89.

The last phase in all this is the display. It's all well and good to be doing all this counting, but it isn't much help to us humans unless the system has a method of revealing it (computers, now there's a different story). So, some means of annunciation is necessary, and that normally takes the form of an LED or LCD numerical display.

This can be done with circuitry similar to that in chapter six, or the counters could be designed with BCD output to directly address the decoder/driver sections. There may well be other approaches I'm not familiar with, but whatever method is used, the end result is the same; that is, the current time displayed in real numbers (that is if you have remembered to correctly set the thing).

So, with that long-winded, but hopefully educational explanation out of the way, let's return to our project.

Project #9: Digital Alarm Clock

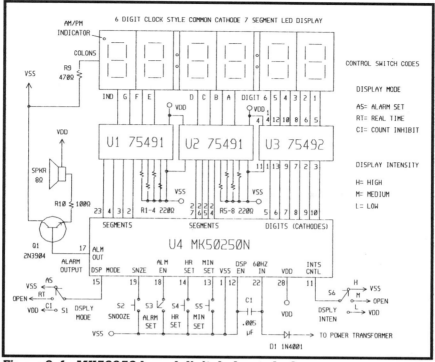

Figure 9-1. MK50250 based digital alarm clock.

For our configuration we need a method to drive the LED display, set the time and alarm, and sound the alarm. Fortunately, all of this is fairly simple. We let the MK50250 do most of the work, and use a 75492 Hex Digit Driver, and two 75491 Quad Segment Drivers to manage the display. These interface easily with the clock chip, and provide ample power for an LED display. Refer to **Figure 9-1** for the complete wiring scheme.

The MK50250 has a 400 to 600 hertz oscillator whose output is designated alarm output, and supplies the tone for the alarm. This is boosted by a single 2N3904 NPN transistor amplifier, and sent to a small 8 ohm speaker.

Three push button switches control the hour and minute set functions, as well as the snooze (10 minute delay) feature. A

SPST switch turns the alarm on and off, while two SP3T rotary or slide switches oversee the display mode and display intensity operations.

The last point of concern is the MK50250 pin 22 connection. This is where the circuit gets its sixty hertz reference signal. As stated before, AC operated clocks use the line frequency as their starting point for division down to one hertz.

CIRCUIT DESCRIPTION

PCB construction is recommended, but with care, point-to-point wiring will also work. Just be careful to observe the hook-up lines. Of course, this project can easily be breadboarded for quick evaluation. See **Figure 9-2** for pin designations.

The layout isn't particularly critical, but as always, sound construction practices not only contribute to "healthy" finished products, but are a good habit to fall into. Since the driver outputs are on both sides of the MK50250, a linear arrangement will probably prove most efficient. However, this is the type of configuration that lends itself well to a variety of layouts.

Once the circuit is assembled, it needs to be set up. Referring to **Figure 9-1**, you will see a total of six switches, all of which control some aspect of the clock's performance. So, let's look at each one, and find out what it does.

Switch S1 is either an SP3T rotary, or SP3T slide switch, and is designated display mode. This allows the MK50250's pin 15 to be connected to either the positive voltage, or ground, or left floating. At +5 volts, the alarm time can be set. At ground, the display freezes, and when floating, the clock displays real time.

S2 is a momentary contact push button switch, and it activates the snooze feature. This halts an active alarm, and delays its next activation for ten minutes.

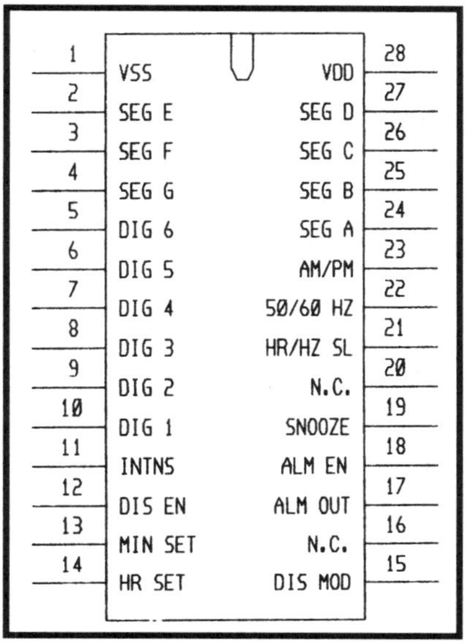

Project #9: Digital Alarm Clock

```
 1  ─ VSS      VDD  ─ 28
 2  ─ SEG E    SEG D ─ 27
 3  ─ SEG F    SEG C ─ 26
 4  ─ SEG G    SEG B ─ 25
 5  ─ DIG 6    SEG A ─ 24
 6  ─ DIG 5    AM/PM ─ 23
 7  ─ DIG 4   50/60 HZ ─ 22
 8  ─ DIG 3   HR/HZ SL ─ 21
 9  ─ DIG 2    N.C.  ─ 20
10  ─ DIG 1    SNOOZE ─ 19
11  ─ INTNS    ALM EN ─ 18
12  ─ DIS EN   ALM OUT ─ 17
13  ─ MIN SET  N.C.  ─ 16
14  ─ HR SET   DIS MOD ─ 15
```

PIN 1	VSS (positive voltage)
PIN 2	Display Segment E
PIN 3	Display Segment F
PIN 4	Display Segment G
PIN 5	Display Digit 6
PIN 6	Display Digit 5
PIN 7	Display Digit 4
PIN 8	Display Digit 3
PIN 9	Display Digit 2
PIN 10	Display Digit 1
PIN 11	Display Intensity Control
PIN 12	Display Enable
PIN 13	Minute Set
PIN 14	Hour Set
PIN 15	Display Mode
PIN 16	Not Connected
PIN 17	Alarm Output
PIN 18	Alarm Enable
PIN 19	Snooze Control
PIN 21	Hour/Hertz Select
PIN 22	50/60 Hertz Input
PIN 23	AM/PM Display Indicator
PIN 24	Display Segment A
PIN 25	Display Segment B
PIN 26	Display Segment C
PIN 27	Display Segment D
PIN 28	VDD (ground)

Figure 9-2. Mostek MK5025 digital alarm clock pinout.

The alarm set switch, S3, is of the SPST variety, and is used to turn the alarm function on and off. This is used to stop the alarm if the snooze option is not desired.

Switches S4, and S5 are both momentary push buttons, and they control the hour, and minute sets, respectively. Holding the button down will rapidly advance the display, while short activations will advance the reading slowly. In either case, the display will advance in an upward direction.

S6 is a second SP3T rotary or slide switch, and controls the display intensity. Connecting pin 11 high will produce the brightest effect, while floating renders a medium intensity, and low provides the dimmest illumination (that's dimmest, not dumbest).

Finally, a few other features bear mentioning. First, the clock normally functions in the 12 hour mode with standard 60 hertz line frequencies. However, if operated on a 50 hertz system, such as European power lines, the clock will function in the 24 hour mode.

Secondly, the AM/PM display indicator will flash whenever a brown out, or power loss, has occurred. This condition prevails until the clock is reset.

Last, if the full six digit display is not used (D1 and D2 do not address seconds, and tens or seconds displays), D1 can be used to flash the colon. This will give an instant indication that the clock is counting. Otherwise, you will have to wait a full minute for the clock to advance one count, indicating that it is, indeed, running.

CONCLUSION

While the MK50250 doesn't provide all the secrets of digital time-keeping, it does offer an insight into this fascinating area of electronics. If nothing else, it will furnish a better understanding of how digital clocks operate.

So, next time you need one of these beauties for the den, bedroom, or elsewhere, buy this chip instead of a factory built job. You can personalize its appearance in any fashion you wish. Just think of the pride you will feel when someone asks where you bought that great looking clock, and you can say, "I built it myself!"

Photo 9-1.
A close-up view of the MK50250 chip. This 28 pin IC is the heart of the digital clock.

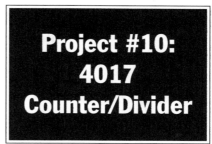

What You Will Need:

Sequential Flasher:
- ✔ U1   LM555, Timer/Oscillator
- ✔ U2   CD4017, Decade Counter/Divider
- ✔ LED1 - 10   Jumbo T-1 3/4 Light Emitting Diodes (any color)

All resistors 5% carbon film.
- ✔ R1   1,000,000 Ohm Potentiometer
- ✔ R2   1,000 Ohm, 1/4 Watt Resistor
- ✔ R3   470 Ohm, 1/4 Watt Resistor
- ✔ C1   2.2 Microfarad Electrolytic Capacitor

Random Number Generator:
- ✔ U1   LM555, Timer/Oscillator
- ✔ U2   CD4017, Decade Counter/Divider
- ✔ LED1 -10   Jumbo T-1 3/4 Light Emitting Diodes (any color)

All resistors 5% carbon film.
- ✔ R1   1,000,000 Ohm Potentiometer
- ✔ R2   1,000 Ohm, 1/4 Watt Resistor

Integrated Circuit Projects

- ✔ R3 — 470 Ohm, 1/4 Watt Resistor
- ✔ C1 — 0.1 Microfarad Mylar Capacitor
- ✔ Sa/b — Momentary Contact, Normally Open Push Button Switch

Frequency Divider:

- ✔ U1 — CD4017, Decade Counter/Divider
- ✔ S1 — SP11T Rotary Switch

In the area of ICs, there are certain ICs that have become Industry Standards. These are the work horses like the LM555, the LM741, the 7490, 7447, and the subject of this chapter; the CD4017 Decade Counter/Divider. This device, sometimes referred to as the "Johnson Counter", is one of the most practical chips ever created. No, I don't know why it's sometimes called that, but a safe guess would be that somebody named Johnson had something to do with its development.

Regardless of who is responsible, this is one IC you will learn to love. The 4017 can be used as either a counter or divider, and employed in a variety of useful and pragmatic configurations. In this chapter we will explore this unique Complementary Metal Oxide Semiconductor (CMOS) IC, and present some circuits that will demonstrate its ability.

IC OVERVIEW

The 4017 comes in a standard sixteen DIP package, usually of the plastic variety (see **Figure 10-1**). The first eleven pins, excluding pin 8 (ground), are the outputs 0 to 9, but they do not

Project #10: 4017 Counter/Divider

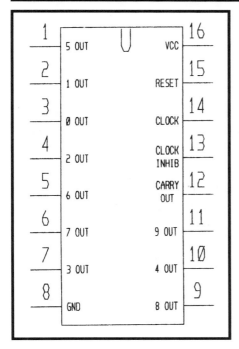

Figure 10-1. CD4017 decade counter/divider pinout.

follow a sequential order. That is, pin 1 is not the zero output, but the fifth output, while pin 2 is the first output. Pin 3, however, is the zero output, and so the designations go, with no apparent order in mind.

This is more of an inconvenience than anything else. It does not affect the chip's operation in any way, but at times does make the circuit board layout a little tricky. Considering all this chip will do, the inconvenience is more than acceptable.

These outputs go high (positive) progressively, starting with zero and ending with nine. At that point, the system starts the cycle over again at zero. This scanning effect runs at the same speed as the clock input, and will continue as long as the clock signal is present.

Alright, with the outputs taken care of, let's look at the next four pins. The first, pin 12, is the carry out line, and this allows for cascading more than one stage. Pin 12 will be connected to pin 14 of the subsequent 4017.

The next line, pin 13, is the clock inhibit. When high, this function prevents the clock signal from reaching the counter, and at times is quite handy.

Pin 14 handles the clock signal which can come from a variety of different sources. In fact, any square wave generator will serve very nicely. More will be said about this in the next section.

The last pin, 15, is the reset, and it performs as all resets perform. With this chip, if taken high, the counter will return to zero and start at the beginning.

Pin 16 is Vcc, or the positive voltage, and is used in conjunction with the ground, pin 8, to provide power for operation. Since the 4017 is a CMOS device, it can handle voltages from three to eighteen volts DC.

CHIP APPLICATION

Having covered the chip description, let's take a glance at some practical circuits. The three presented here should give you an idea of how to put the 4017 to use as a counter or divider.

Figure 10-2 illustrates the configuration probably most often employed for this chip. This is known as a sequential flasher, and is handy for all sorts of purposes. In this circuit, ten LEDs are consecutively illuminated by the 4017 outputs.

The clock signal is provided by our old friend the LM555 timer/oscillator arranged as an astable oscillator. This pulse, adjustable with potentiometer R1, sets the pace, or speed of the sequence. That speed can be a very slow movement, or can be increased so that all ten LEDs appear to be on at once.

In reality, only one of the LEDs is actually on, but due to persistence of vision, the eye and brain perceive all indicators to be

Project #10: 4017 Counter/Divider

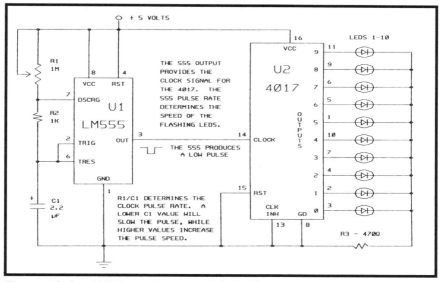

Figure 10-2. 4017 based sequential flasher.

on. Persistence of vision is a useful and interesting attribute of human sight. Research on this subject can provide some extensive but fascinating reading, not to mention time consuming. However, I will try to give a simple explanation of this characteristic.

The eye has the ability to retain anything it sees for about 1/20 of a second (when I say eye I am talking about a combination of the eye and brain), so any series of images that appears at a faster rate than 20 hertz will overlap. This is due to the eye not having erased the first image before the second one comes along.

Thus, when an LED is flashing on and off at 20 hertz or faster, it will appear to be on constantly. The same principle applies to motion picture and video frames that make up the films/videos we watch. In each case, the film is made up of thousands of individual still images, but since they are being presented to the eye at a rate higher than 20 hertz (usually 24 frames a second for normal speed), they overlap, giving the appearance of motion.

Integrated Circuit Projects

What all this means is that if you want to observe the sequential movement of the 4017's LED display, you will need to set the clock speed at less than 20 hertz. In most cases, somewhere between 3 and 10 hertz is about right. It just depends on the effect, or result you are shooting for.

This circuit has all kinds of possibilities in games, attention getting displays, advertising, and other applications where a moving row of LEDs is desirable and/or functional.

In **Figure 10-3**, you will see a circuit almost identical to the one in **Figure 10-2**. This configuration results in a random number generator, and the primary difference is the addition of a momentary contact, normally open push button switch, Sa/b. I have illustrated the switch in two different locations, but only one is required. The purpose here is to connect or disconnect the clock as needed.

With switch Sa, the voltage path is interrupted, thus disabling the clock, while Sb simply prevents the clock's output from reaching the 4017 input. Either will work, it's your decision.

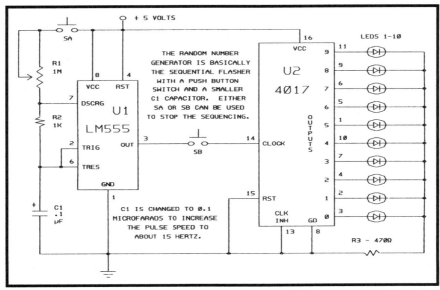

Figure 10-3. *4017 based random number generator.*

Project #10: 4017 Counter/Divider

As for operation, you will want to set the clock speed at well above the 20 hertz rate. That will keep the full display on (at least to the eye) as long as Sa/b is pressed, and when released, the display will freeze on one of the ten LEDs.

The high oscillator rate will obviate the chances of someone being able to outguess the sequence, thus halting the display on a particular LED. In reality, this is not a true, absolute random number generator, but for our purposes, such as a game, it will do right nicely.

Of course, each of the LEDs will have to be numbered from 1 to 10, or 10 to 100, or whatever, in order to provide a random value. Another version would be to employ only the first six stages, leaving the last four outputs floating (open). In this fashion, the random number would be between one and six, and could be substituted for a die in board games.

The last circuit, **Figure 10-4**, portrays the 4017 in the role of a frequency divider. By connecting one of the outputs to the reset you can divide the frequency of the clock input by anything from "0" (no change) to "10". The new frequency (output) is available at the "carry out" line, pin 12.

For example, in the last chapter I talked about the MM5369 chip producing a 60 hertz output that could be divided down to 1 hertz. Referring to **Figure 10-4**, it is easy to see how this can be done.

If you arrange the first 4017 as a divide-by-six counter (output 6, pin 5 to reset, pin 15), and apply the 60 hertz to pin 14, the output at pin 12 will be 10 hertz. By setting up the second 4017 as divide-by-ten (reset to ground), the 10 hertz signal will be divided down to 1 hertz. Now you have your "seconds" count signal.

This same principle will work with any frequency to about 20 megahertz. Merely connect reset to the output pin that corre-

Integrated Circuit Projects

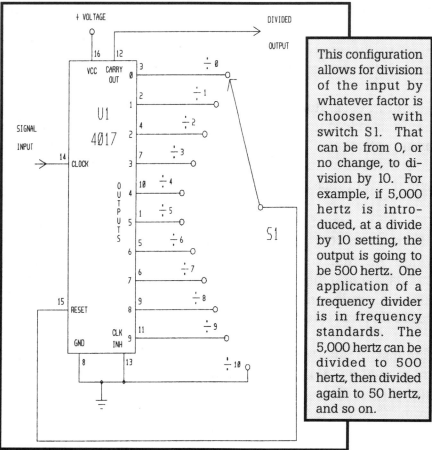

Figure 10-4. 4017 based frequency divider.

This configuration allows for division of the input by whatever factor is choosen with switch S1. That can be from 0, or no change, to division by 10. For example, if 5,000 hertz is introduced, at a divide by 10 setting, the output is going to be 500 hertz. One application of a frequency divider is in frequency standards. The 5,000 hertz can be divided to 500 hertz, then divided again to 50 hertz, and so on.

sponds to the desired division, and voila, you have the new frequency at pin 12. As many stages as necessary can be cascaded together, so theoretically you could take frequencies in the megahertz down to single pulses, or even 10, 100, or 1,000 second time periods.

CONCLUSION

This chapter has presented the fundamentals of utilizing the CD4017, but in no way has it covered the chip's total capability.

Project #10: 4017 Counter/Divider

However, if you have comprehended what was set forth here, you will be able to expand on these basics, and find numerous application for the Johnson counter.

As stated at the outset, there are Industry Standards, and this chip truly qualifies as one. That makes it a treat to use. It is extremely nice to work with an IC that is so utterly pragmatic. In the end, the 4017's abilities are such that you will be amazed how often it fits the bill for one of your prototypes.

If I seem to be heaping immense praise on this device, it is because of the admiration I have developed for this ICs design and utility. After using it a couple of times, I think you, too, will recognize its attributes.

Photo 10-1. Breadboard layout of the sequential flasher circuit.

Integrated Circuit Projects

Project #11: FM Wireless Microphone

What You Will Need:

Wireless Microphone:
- ✔ Q1NPN — 2N3904 NPN Silicon Transistor
- ✔ Q1PNP — 2N3906 PNP Silicon Transistor

All resistors 5% carbon film.
- ✔ R1 — 100,000 Ohm, 1/4 Watt Resistor
- ✔ R2 — 47,000 Ohm, 1/4 Watt Resistor
- ✔ R3 — 470 Ohm, 1/4 Watt Resistor
- ✔ Rm — 1,000 Ohm, 1/4 Watt Resistor (optional Electret microphone)
- ✔ C1 — 2.2 Microfarad Electrolytic Capacitor
- ✔ C2 — 470 Picofarad Ceramic Disk Capacitor
- ✔ C3 — 5 to 30 Picofarad Trimmer Capacitor
- ✔ C4 — 5 Picofarad Ceramic Disk Capacitor
- ✔ L1 — 5 to 6 turns, #22 Bus Wire, 1/4 Inch Diameter, Spread 6 to 1/2 Inch
- ✔ MIC — Electret Microphone (optional – see text)
- ✔ ANT — Up to 24 Inch Wire, Rod, or Telescoping Antenna

Integrated Circuit Projects

Audio Preamp FM Microphone:

- ✔ Q1, 2 — 2N3904 NPN Silicon Transistors

  All resistors 5% carbon film.
- ✔ R1 — 1,000 Ohm, 1/4 Watt Resistor
- ✔ R2 — 330,000 Ohm, 1/4 Watt Resistor
- ✔ R3 — 15,000 Ohm, 1/4 Watt Resistor
- ✔ R4 — 100,000 Ohm, 1/4 Watt Resistor
- ✔ R6 — 470 Ohm, 1/4 Watt Resistor
- ✔ C2 — 0.001 Microfarad Electrolytic Mylar Capacitor
- ✔ C3 — 2.2 Microfarad Electrolytic Capacitor
- ✔ C3 — 470 Picofarad Ceramic Disk Capacitor
- ✔ C5 — 5 to 30 Picofarad Trimmer Capacitor
- ✔ C6 — 5 Picofarad Ceramic Disk Capacitor
- ✔ MIC — Electret Microphone
- ✔ L1 — 5 to 6 Turns, #22 Bus Wire, 1/4 Inch Diameter, 1/2 Inch Spread
- ✔ ANT — Up to 24 Inch Wire, Rod or Telescoping Antenna

Note:

Any of these circuits can be operated off battery power from 6 to 18 volts, or from an AC wall socket style power supply. If the AC system is used, be sure the supply is properly filtered to prevent an annoying 60 hertz hum.

This time around we are going to examine an all time favorite, the wireless microphone. Whenever this subject comes up, I fondly remember my own introduction to electronics. It was an Allied Radio, Knight Kit experimental lab. You know, the kind where you use the same parts to build thirty five, or fifty, or however many different electronic circuits.

I look back on that kit, and while I thoroughly enjoyed each of the separate projects, the one that made the biggest impression was the wireless microphone. We are talking many moons ago (Knight Kits have been defunct since Ponce De Leon was searching for the Fountain of Youth, or so it seems), thus, this transmitter operated on the AM broadcast band. But that didn't matter. The ability to send my voice or music to a radio was pure fascination.

For that matter, the fascination continues to this day. I have always enjoyed working with oscillators and transmitters. So, if you share my sentiments, you will relish this chapter. In it, we will explore the world of the oscillator, and build a transmitter that transmits to the FM broadcast band. Whether you use it as a neighborhood broadcast station, or just to send music from the den to your workshop, this project will illustrate the principle behind transmitters, and do so with a working device.

CIRCUIT THEORY

Referring to **Figure 11-1**, you will see two separate circuits. One uses a NPN transistor, while the other employs a PNP transistor. The purpose of this illustration is to point out that an oscillator can be constructed from either type of semiconductor. There are differences in the circuit configuration, though.

First, let's look at why the transistor is defined as either NPN or PNP. This has to do with its architecture in that an NPN transistor is charged negatively, and a PNP is charged positively. That is a very simplistic explanation, but to go further requires getting into the actual manufacturing techniques, and that is complicated and somewhat confusing. Besides, I'm not sure I totally understand it myself.

So, looking at the schematics, you will notice that with the NPN configuration the emitter is connected to ground (through

Integrated Circuit Projects

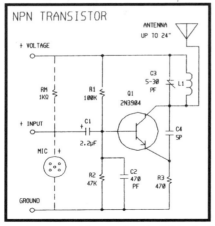
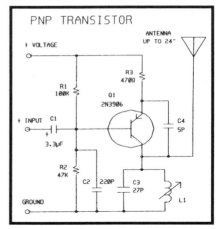

Figure 11-1. FM wireless microphone circuits.

In this circuit an NPN transistor is used as the oscillator. The connections to the electret microphone and resistor RM are shown as dashing lines due to the optional status of these components. The wireless mic can operate as a stand alone system. Audio sensitivity will not be as good with a preamplifier.

Again, the mic and RM can be added if desired. However, performance will be better with an amplified input. Voltage for either circuit can range from 3 volts to 15 volts, with the higher voltages improving the transmission range. Also, a fixed coil and variable capacitor could be employed with this circuit.

resistor R3), and the PNP circuit connects the emitter to the positive voltage (again through R3). Additionally, the "tank" circuits (C3/L1) are reversed in their power connections.

Other than that, the two schematics are pretty much the same. Which type of transistor selected has little to do with the overall performance of the oscillator. Incidentally, this is what is referred to as a modified Hartley oscillator.

With all that in mind, let's examine what each component does. Of course, in this arrangement the transistor acts as an oscillator instead of a switch. The specific type of semiconductor used (2N3904, 2N2222, 2N4401, etc.) governs the maximum fre-

Project #11: FM Wireless Microphone

quency at which the transmitter can operate, thus the 2N3904 and 2N3906 were chosen as they will reach to between 250 to 300 megahertz.

Resistors R1 and R2 form a voltage divider that provides bias for the transistor, while C2 furnishes high frequency by-pass. Capacitor C1 is for input coupling, and is only needed when an audio signal is introduced to the transmitter.

As earlier mentioned, capacitor C3 and coil L1 form the tuning, or tank circuit, and either one can be variable. You'll note that the NPN oscillator uses an adjustable capacitor, while the PNP circuit adjusts the coil. In either case, this controls the operating frequency.

The last two components for the basic oscillator are C4, the feedback capacitor, and R3 the current bias resistor. C4 helps the circuit maintain oscillation, and R3 controls the current through the transistor, thus controlling the output power.

So, as seen, the fundamental circuit is not at all complicated. If desired, an Electret microphone can be connected at the input (as illustrated in NPN schematic), and this will provide reasonably sensitive audio pickup. The Electret style microphones have a built in field effect transistor (FET) pre-amp which provides some input gain. However, they do require a positive voltage connection. That is the purpose of resistor Rm connected to the (+) voltage rail.

Figure 11-2 depicts a configuration that will increase the audio sensitivity a great deal. Here, we use a single transistor pre-amp to boost the input to the oscillator stage. Again, a 2N3904 transistor is chosen with C1 providing coupling, C2 acting as by-pass, and resistors R2 and R3 furnishing circuit bias. R1, as Rm in the previous circuit, powers the Electret microphone.

Integrated Circuit Projects

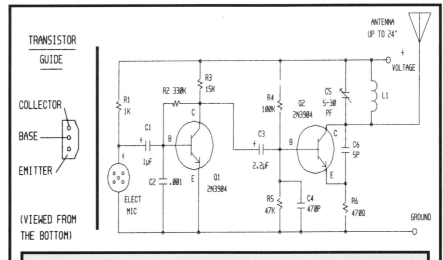

This is the NPN transistor oscillator with the addition of a preamplifier section. The amp vastly improved the audio sensitivity of the transmitter to a point where a whisper can be heard from several feet away. The voltage can be from 3 volts to as high as 15 volts, with a 9 volt battery doing a good job. The overall range, or transmission distance, is somewhat controlled by the power supply voltage. The higher that potential, the farther the transmitter will reach. The second major factor in range is the antenna length. 24 inches is the legal limit, and with the antenna at that length and the battery in the 9 to 12 volt range, indoor distances of 300 to 400 feet can be expected, with outdoor ranges in excess of 1000 feet.

Figure 11-2. Audio preamp FM microphone. L1- 6 turns of #22 bus wire on a 1/4" coil form (wooden dowel).

CIRCUIT DESCRIPTION

Now that we all understand how the oscillator works, let's get down to the construction details. This is the best part, anyway. First, the tank circuit. This is comprised of C1 and L1, and neither component is too terribly critical. If you use a variable capacitor, then coil L1 will be 5 or 6 turns of number 22 bus wire wound on a 1/4 inch form (dowel, drill bit, pencil, etc.). Spread out the turns to a length of about 1/2 inch.

Actually, the number of turns is not as important as the total length of the wire. That needs to be six to eight inches, thus you could wind the coil on a smaller diameter form, resulting

Project #11: FM Wireless Microphone

in more than 6 turns. However, performance, especially range, seems to decrease slightly with smaller diameter coils, so you may want to stay with the 1/4 inch size. Be sure that the coil windings do not touch.

With the second option, a variable coil, you will want to wind the same six to eight inches of number twenty-two wire on a slug tuned 5/16 inch coil form. The form can be robbed from an old IF transformer, or other adjustable RF coil. In this case, the wire needs to be of the enameled, or insulated variety, as the turns will come in contact with each other. Usually, the coil will result in 4 to 5 turns.

As for the capacitor, in the first scenario it will be a 5 to 30 picofarad variable trimmer, and for the second system, a 15 to 33 picofarad fixed disk capacitor. The value of the tank capacitor will determine the area of the FM band in which the transmitter will operate. For example, with the NPN circuit, 5 picofarads will put the signal near the top of the band, while 30 picofarads drops it down into the band's initial frequencies.

Although the two tuning circuits are shown in specific configurations, they are interchangeable. The NPN transmitter could use a variable coil system, and the adjustable capacitor approach could be employed in the PNP oscillator.

Concerning construction method, as usual either PCB, or point-to-point can apply. PCB will probably result in a more stable and durable arrangement, and the board can be smaller than one inch square, if space is a consideration. However, either method is more than acceptable.

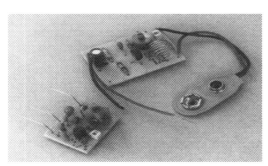

Photo 11-1. Two different arrangements for the basic FM wireless microphone.

To stay within Federal Communications Commission (FCC) regulations, the antenna length should not exceed twenty four inches. Additionally, the power output must stay within certain limits. These limitations are expressed in part 15 of the FCC rules and regulations, and while no specific milliwatt limit is placed on FM band transmitters, it does specify the millivolt reading on a field strength meter, at ten meters, and so forth.

These are difficult readings to take unless you possess some fairly fancy equipment, but if the schematic values are observed, you will stay well within the limits, and legal. Lowering the value of the current bias resistor (R3) will increase the transmitter's output, but you have to be careful not to exceed FCC allowances.

The input can come from several different sources. As seen in **Figure 11-1**, an Electret microphone will provide acceptable results, and when connected to a separate pre-amp (**Figure 11-2**), will produce excellent results (in regards to sound sensitivity).

A crystal microphone will also suffice in the **Figure 11-1** circuits, but the sensitivity will not be as good. In addition, the sound will have a tinny quality due to the characteristics of this type of microphone.

Dynamic pickups will improve sound quality, but will definitely require the pre-amp stage shown in **Figure 11-2**. Their sensitivity is good, however both dynamic and crystal microphones are larger in size than the Electret variety.

Another good input source is from the amplifier stage of a tape player, CD player, radio, PA amp, or any other piece of equipment that has such an output. This is a great way to send your favorite recordings to other areas of your home or office. It won't be in stereo, but it is cheaper than buying a second stereo system.

Project #11: FM Wireless Microphone

Using the amplifier output approach allows you to expand the system if you decide to go the DJ route. Employing a mixer, a tape player can be one channel, a CD another, the voice microphone a third, and so forth. The mixer will provide balancing and intensity for the various input sources, thus music, voice and any other desired information can be simultaneously transmitted by the oscillator.

The last consideration is the power source. A 9 volt battery will operate the circuits in both **Figure 11-1** and **Figure 11-2** for short periods of time (approximately 4 to 5 hours continuously), but for extended use, either a larger battery or an AC based supply will be needed.

If the AC strategy is taken, be sure the power supply is well filtered (large filter capacitors). This will help prevent the oscillator from picking up and transmitting the 60 hertz line frequency, or hum.

CONCLUSION

OK, Goatman Jack, or Jackie, you're ready for your broadcast debut. Scare up your favorite tunes, and hit the air a runnin'. Be sure and let the neighbors know the time and channel, however. Of course, if you plan to use the transmitter for another purpose, this, too, is the time to try it out.

With power connected, find a dead spot on the dial, adjust the variable capacitor, or coil, until you hear your signal, and you're ready to go. On the average, with a favorable location (as high up as possible, second story, attic, etc.), these circuits will transmit a quarter of a mile or better, and be perfectly legal.

Who knows, this may be opportunity knockin'. A big, high power radio executive might pass through your neighborhood one day, hear your golden voice on the air, and sign you to a multi-million dollar contract.

Integrated Circuit Projects

What You Will Need:

- ✔ U1 — LM356N, JFET Input Op-Amp, Wide Band
- ✔ U2 — LM3914, Bargraph Display Driver
- ✔ D1 — 1N34 Germanium Detector Diode
- ✔ D2 — 1N4001 Rectifier Diode
- ✔ LED — Jumbo T-1 3/4 Light Emitting Diode (green)
- ✔ DSPLY — 10 Element Bar Style LED Display (Electronic Goldmine, Jameco)

All resistors 5% carbon film.

- ✔ R1, 6 — 470 Ohm, 1/4 Watt Resistors
- ✔ R2 — 5,000 Ohm Potentiometer
- ✔ R3 — 1,000,000 Ohm, 1/4 Watt Resistor
- ✔ R4 — 4,700 Ohm, 1/4 Watt Resistor
- ✔ R5 — 10,000 Ohm Potentiometer
- ✔ R7 — 1,000 Ohm, 1/4 Watt Resistor
- ✔ R8 — 15 Ohm, 1/4 Watt Resistor
- ✔ C1 — 500 Picofarad Ceramic Disk Capacitor
- ✔ C2, 3 — 0.02 Microfarad Mylar Capacitors
- ✔ C4 — 47 Microfarad Electrolytic Capacitor

Notes:

This unit operates well off a standard 9 volt alkaline battery. The antenna can be a wire or rod, however a telescoping style whip of 30 to 40 inches will provide more portability.

Unfortunately, in this day and age of sneaky people (not to mention the crooks), we all have to protect our privacy whether we want to or not. This is especially true if you own a business, and have secrets from your competition. I hate to sound like the voice of doom, or the novel <u>1984</u>, but a harsh reality is that it's all too easy for someone to plant a listening device, or bug in your midst without your even suspecting its presence.

For this reason, it never hurts to sweep your home and/or office from time to time just to be sure it is secure. With the device we will explore in this chapter, that becomes a simple matter. The Budget Bug Detector is easy and inexpensive to build, but of greater importance, it will do wonders for your peace of mind.

If there is a culprit out there that has invaded your solitude, this device will alert you. You may want to extract the bug and smash it with a hammer, or leave it where it is and pass on, as the intelligence folks would say, disinformation (bad stuff). That might just teach your intruder a nasty lesson.

CIRCUIT THEORY

The bug detector is a specialized RF receiver based on what is commonly referred to as a passive detector. That term comes

Project #12: RF Bug Detector

from the absence of an oscillator section used to detect the RF signals. This is somewhat important in a bug detector because those oscillators do produce a signal of their own, and you don't want the detector detecting itself.

Referring to **Figure 12-1**, the antenna picks up the energy from a transmitter, and sends it to the L1/D1 network. Here it is rectified, then sent to a wide band op-amp, U1. The op-amp boosts this signal before applying it to the circuit's display section.

Resistors R5 and R8 form a voltage divider used to adjust the sensitivity of the receiver, while R2 and R3 comprise a second voltage divider controlling the op-amp gain. Between the two,

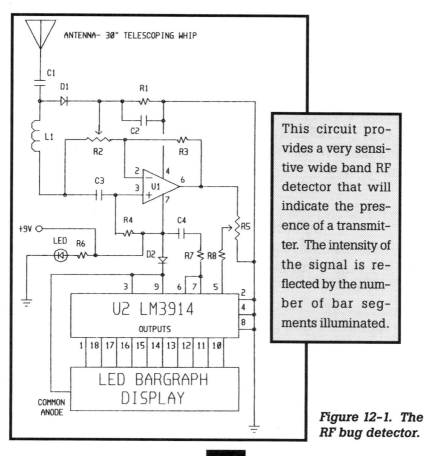

This circuit provides a very sensitive wide band RF detector that will indicate the presence of a transmitter. The intensity of the signal is reflected by the number of bar segments illuminated.

Figure 12-1. The RF bug detector.

a lot of latitude is provided in terms of detector response. For most applications, you will want maximum sensitivity, but there are times, when dealing with high power transmitters, that you will need to back it off a bit.

Displaying the results of the detector is handled by our friend the LM3914 Bargraph Display Driver; remember chapter one? As the strength of the signal increases, the voltage output of the op-amp also increases. This, in turn, activates more of the LM3914 outputs thus lighting additional display LEDs.

So, as can be seen, this is not a complicated piece of equipment. The passive detector recognizes RF at virtually any frequency (I've received responses off a radar gun), and it spots signals from even the microwatt bugs. However, its efficiency depends largely on its application, and this will be covered in depth later on.

CIRCUIT DESCRIPTION

The prototype uses PCB construction, and is housed in a small project case (1x3x5 inches). It could be smaller, but these dimensions make it pocket sized, thus very portable. Additionally, it fits nicely in the hand, affording a very functional design.

However, the configuration lends itself well to a variety of enclosures, so this allows you to choose. That also applies to the wiring technique, but keep it clean, especially the L1/D1 detector section.

Potentiometers R2 and R5 can be panel mounted to allow for easy adjustment, but unless you plan on checking for high power transmitters on a regular basis, it is probably better to use board mounted components. Once set, they are not often changed, and access holes can be drilled to provide a means of adjusting these controls.

Project #12: RF Bug Detector

With the circuit completed, it needs to be adjusted, or calibrated. The first step is to set the op-amp gain with R2, and a good starting point is mid scale. This may need additional adjustment after you have set the sensitivity, but you don't want to overdrive the op-amp.

Now, place the detector in an area that is at least five feet from any source of RF energy. This would include television sets, radio receivers, computers, and, of course, any type of transmitter. R5 is then set to just turn on the first display LED. Next, back the potentiometer off to just turn off the LED. If you encounter any difficulty with this, try slightly readjusting R2.

This procedure sets the detector for maximum sensitivity, and it is now time to test it. Employing a source of RF, say, one of the oscillators in the last chapter, a cordless telephone, a walkie-talkie, etc., bring the detector towards the source. If all is working, the closer you get, the more display elements light up, until the entire display is on.

With the FM wireless microphone, full display should occur with the detector's antenna about a foot from the transmitter's antenna. A little farther distance, maybe a foot and a half, or two feet, will be experienced with a cordless phone, or small walkie-talkie. A large 5 watt transceiver (CB variety) will pin the display at several feet, as will a base station. Amateur radio transmitters, depending on their output power, can achieve a full scale reading at up to a hundred feet.

Hence, the oscillator's power plays a significant role in the detector's response. However, since most planted bugs are going to be of a low power design, the maximum sensitivity setting will serve you well.

Higher output bugs do exist, but they require more power to operate, and are usually difficult to install. Don't let that lull you into a false sense of security, since the standard AC wiring is convenient, a connection can be made, and there is little chance you would notice it, even on your power bill. The advantage here is a permanent bug, not disabled when the battery goes dead.

A couple of comments about sweeping a room, office, or other area. There are some key words and phrases involved in this, such as Patience, thoroughness and attention to detail. Adhering to all will insure a precise and complete search.

With a room, start at the entrance and move around each wall until you are back at the entrance. It doesn't matter in which direction you go, but be sure to check from floor to ceiling all the way around. Next, systematically move from the walls to the center of the room, again examining top to bottom.

Bugs are easily hidden in furniture, lamps, plants, books, desk sets and basically anything else that will conceal them. So, be thorough, and don't miss those subtle locations. Without patience, and attention to small details, you are probably defeating the purpose of the sweep. One commonly employed trick is to plant one bug practically in plain sight, while another is buried, thus overlooked by a sloppy sweep.

The procedure for offices and larger areas is pretty much the same as for a single room. Naturally, the more extensive the area, the more tenacious your inspection has to be. Ultimately, by following the guidelines, you can accomplish a comprehensive search and spot any RF listeners if they are present.

Once a signal is detected, carefully move in on it to establish its exact location. As earlier stated, it is often to your advantage to leave a bug in place, so be cautious not to reveal its discovery. A normal conversation taking place in the room being searched will help cover the sweep activity, thus keeping the bug's detection confidential.

CONCLUSION

With the Budget Bug Detector in hand, you have equipped yourself with one more weapon against the unscrupulous elements of our society. It is sad the practice of illegal bugging has become so widespread, but that doesn't mean you have to fall victim to it. A periodic screening of your home and office can prevent such unauthorized and aggravating intrusions.

There are, of course, other types of bugs around that do not use radio waves to communicate their sinister eavesdropping, but by far, the most commonly employed devices are wireless transmitters. Their ease of use and small size are a natural for this type of activity. Additionally, simple oscillator circuits, more than adequate for the purpose, abound in electronics literature (chapter eleven in this text, for example).

All of this does allow for unprincipled, illegal and irresponsible application of these devices by a certain segment of our culture. This is not to say that electronic surveillance isn't appropriate in areas of law enforcement, national security or the legal protection of life and property, but it is to say that too often the surveillance is used for criminal and self-serving purposes.

Photo 12-1. The completed bug detector.

Integrated Circuit Projects

Project #13: Passive Detector Receiver

What You Will Need:

- ✔ U1 — LM386N-3, Low Voltage Audio Amp 500mw/9V
- ✔ Q1 — 2N3904 NPN Silicon Transistor
- ✔ D1, 2 — 1N34 Germanium Signal Diodes

All resistors 5% carbon film.

- ✔ R1 — 330,000 Ohm, 1/4 Watt Resistor
- ✔ R2 — 15,000 Ohm, 1/4 Watt Resistor
- ✔ R3 — 10,000 Ohm Potentiometer
- ✔ R4 — 10 Ohm, 1/4 Watt Resistor
- ✔ C1 — 0.005 Microfarad Ceramic Disk Capacitor
- ✔ C2 — 1 Microfarad Electrolytic Capacitor
- ✔ C3 — 0.001 Microfarad Ceramic Disk Capacitor
- ✔ C4 — 10 Microfarad Electrolytic Capacitor
- ✔ C5 — 100 Microfarad Electrolytic Capacitor
- ✔ C6 — 220 Microfarad Electrolytic Capacitor
- ✔ C7 — 0.05 Microfarad Ceramic Disk or Mylar Capacitor
- ✔ T1 — Audio Transformer (8 to 1.2K ohm ratio): XFR1028A14 or Equivalent (B.G. Micro)

Integrated Circuit Projects

✔	S1	SPST Switch (may be included on volume control)
✔	J1	Phone Jack (either open or closed circuit variety)
✔	B1	Battery (9 volt alkaline recommended)
✔	ANT	Several Hundred Turns of #30 or #32 Enameled Wire (see text)

This receiver is safe to use around communications equipment. Safe in the sense that it will not interfere with the normal operation of said equipment. How could a receiver interfere, you ask? Well, remember from the last chapter when I talked about receivers using oscillators to detect incoming signals. Those oscillators generate a weak signal of their own, and it is this signal that can cause the trouble.

What is the solution? A passive detector. So, here we have a second receiver that employs a non-oscillator passive detector, and that makes it neutral to the equipment it is monitoring. In terms of practicality, most airlines will not allow you to operate a conventional (superheterodyne) radio during a flight. This is due to their fear of interference with the aircraft's electronic gear. However, this device is perfectly safe, since without an oscillator it can not possibly cause a problem.

Now, when flying (in airplanes) you can listen in on the crew's radio transmissions, and when close enough, the tower as well. Best of all, there is no danger of accidentally sabotaging the craft's ability to communicate, or disrupting its radio navigational equipment. After all, you don't want to end up in Boston when you were headed for New York (I don't know, maybe you do).

Well anyway, the airline would take a dim view of it, thus it is best to be safe and certain. But that is not the only application

Project #13: Passive Detector Receiver

for this receiver. No sir. It will also pick up signals over a wide range of frequencies, such as the citizen's band, amateur radio bands, commercial business bands, and when close enough, broadcast bands. Hence, this simple device has all kinds of possibilities.

CIRCUIT THEORY

This configuration is slightly different from the passive detector used in the previous chapter, but essentially, it does the same job. Referring to **Figure 13-1**, you'll notice that two germanium diodes are used instead of just one, and that D2 is connected as the coil was in the other circuit. Actually, the primary winding of the transformer functions as the coil, as well as part of the induction system.

Once a signal has been detected, transformer T1 serves to boost the incoming signal and buffer it from the rest of the circuit. This conditions that signal for the pre-amp/audio amp stage.

The next step is to introduce the signal to a simple single transistor pre-amp which provides the necessary input level for the LM386 (U1). Note that the pre-amp output is connected to potentiometer R3, with the "wiper" (center connection) going to the non-inverting input of U1. The inverting input is connected to ground, as a reference, thus allowing the changing signal from R1 to control the amplifier's input gain and output volume.

C2 is used to increase the amp's overall gain to 200. If C2 is left out, the 386 has a gain of only 20. The series network of C7/R4 provides output stability by offsetting the inductance of the output load (phones, speaker). Without this network, distortion is likely. By providing DC blocking, C6 serves as the coupling capacitor from the LM386 output to the load. The "load" can be either low impedance earphones, or a 16 ohm speaker, which ever is most convenient for your need.

Integrated Circuit Projects

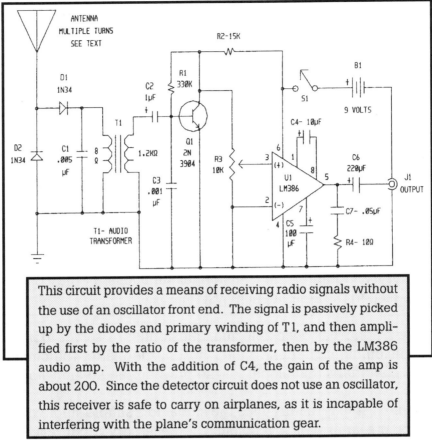

This circuit provides a means of receiving radio signals without the use of an oscillator front end. The signal is passively picked up by the diodes and primary winding of T1, and then amplified first by the ratio of the transformer, then by the LM386 audio amp. With the addition of C4, the gain of the amp is about 200. Since the detector circuit does not use an oscillator, this receiver is safe to carry on airplanes, as it is incapable of interfering with the plane's communication gear.

Figure 13-1. Passive detector receiver.

As for power, any battery arrangement from 6 to 15 volts will suffice, however a standard 9 volt alkaline provides excellent results and compact size. Of course, you can employ an AC style power supply if that better suits your purpose.

CIRCUIT DESCRIPTION

Using PCB construction, this configuration can be assembled in a rather small case, something along the lines of 1x2x3 inches, which makes it highly portable. The major space concern will be the battery.

Project #13: Passive Detector Receiver

You will want the volume control (R1) panel mounted to allow access, as well as the power (On-Off) switch, although it can be incorporated in the potentiometer. If you decide to use headphones, a jack will have to placed in the case to accommodate these. However, a larger size enclosure will provide room for a small speaker, and a closed circuit phone jack can be employed to disable said speaker if headphones, or an earphone is plugged in.

The antenna needs to be as long as possible, as this will affect the sensitivity of the receiver. One method is to use very fine enameled wire (#30, or #32), and wind several hundred turns around the perimeter of the case. Once wound, this antenna "coil" can be removed and fitted to the inside of the case, then secured with epoxy, or silicon cement.

The loose end will be connected to the circuit board, after cleaning off the enamel with sandpaper, or a flame, or both. You will be surprised at the performance this arrangement affords. For the most part, the antenna is omni-directional, but under certain conditions it may display slightly directional characteristics, similar to those of an AM broadcast receiver.

With everything in place, turn on the power, and adjust the volume. The LM386 produces very little noise, so you shouldn't hear that annoying "hiss" associated with many receivers. Be careful, however, not to turn the volume up too high, especially with headphones, as an incoming signal could get dangerously loud.

It goes without saying, there is no tuning involved. Whenever a transmitter in the near vicinity (up to several hundred feet, depending on the transmitter's power) is keyed, the receiver will detect the signal, and permit you to listen in on the audio. Additionally, you will be able to hear Morse code, telex signals, modems, or any other information being broadcast.

CONCLUSION

That is about all there is to it. A simple design that will let you receive a wide range of frequencies over a fairly close proximity. Kind of like an audio field strength meter.

I think you will find numerous applications for this circuit; I have. If nothing else, it is a quick and easy way to check oscillator and transmitter designs, or test transceivers, cordless telephones, RF remote controls and the like for proper operation.

So, this is one handy gadget to have not only on the work bench, but with you. I guarantee that if you don't build this one, you will be grumbling to yourself the next time you board an airliner. That's a limited guarantee, by the way!

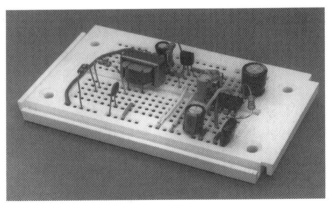

Photo 13-1. Solderless breadboarded version of the passive detector receiver.

Project #14: RCA Voltmeter

What You Will Need:

✔	U1	CA3161E, BCD to 7 Segment Decoder/Driver
✔	U2	CA3162E, Dual Slope & Rate Analog/Digital Converter
✔	Q1 thru 3	2N3904 NPN Silicon Signal Transistors
✔	DSPLY	3 Digit Common Anode Multiplexed LED

All resistors 5% carbon film.

✔	R1	10,000 Ohm, 1/4 Watt Resistor (0-9.9V)
		100,000 Ohm, 1/4 Watt Resistor (99.9V)
✔	R2	1,000 Ohm, 1/4 Watt Resistor
✔	R3	10,000 Ohm Potentiometer
✔	R4	50,000 Ohm Potentiometer
✔	R5	470 Ohm, 1/4 Watt Resistor
✔	C1	10 Microfarad Electrolytic Capacitor
✔	C2	0.33 Microfarad Electrolytic Capacitor

Integrated Circuit Projects

Notes:

This circuit can be operated on batteries, but due to the high current drain of LED displays, AC line power is recommended. It is also recommended that the power supply be regulated to insure consistent readings.

Don't mind the title, it's just the result of a momentary lapse. I'm all right now, honest. What we want to do in this chapter is construct a simple circuit that will display from 0 to 100 volts in three digit resolution. This can be a very useful schematic to have around, as it has virtually endless applications.

Any time you would like to know what the voltage is, this ol' boy can be plugged in and provide that information. Thus, it is not only useful in multimeter capacities, but in a whole spectrum of areas such as digital thermometers, transmitter final stage monitoring, digital test equipment, and power line monitors, just to name a few.

CIRCUIT THEORY

Take a look at **Figure 14-1**, and you will see that this circuit revolves around two ICs; a CA3161E and a CA3162E. Both are manufactured by RCA, and the 3162 is an A/D Converter, while the 3161 is a BCD to 7 Segment Decoder/Driver.

When a voltage is applied to the input, it is an analog quantity, so conversion to a digital quantity is necessary to display the voltage. This is the role of the A/D converter. Through a dual slope arrangement of timing a discharge against a reference, the 3162 is able to produce a binary coded decimal (BCD) equivalent of the input voltage.

Project #14: RCA Voltmeter

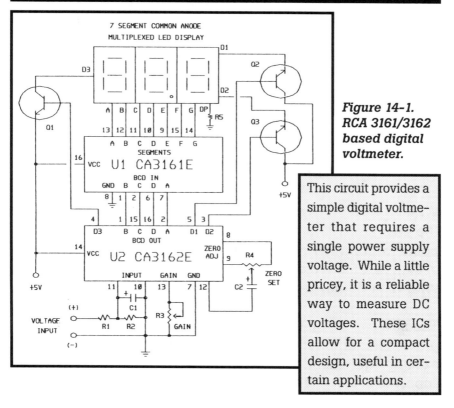

Figure 14-1.
RCA 3161/3162 based digital voltmeter.

This circuit provides a simple digital voltmeter that requires a single power supply voltage. While a little pricey, it is a reliable way to measure DC voltages. These ICs allow for a compact design, useful in certain applications.

Remembering back to the chapters on counters, you'll recall that a BCD is four digits of either high (1) or low (0) values, and each represents an individual number. That code first has to be decoded in order to drive the LED display.

In the counter circuits we used a 74LS47 for that purpose, and it would work here as well. However, since we are using an RCA A/D converter, let's stay with the Radio Corporation of American for the display decoder/driver. The CA3161 is designed for use with the 3162, and although a little more expensive than the 74LS47, it is a good choice.

A second function the CA3162 provides is display multiplexing. It will automatically decide which of the three common anodes needs to be activated, thus insuring a proper display value. NPN signal transistors, Q1-3, are used to boost the current and brightness of the LED segments.

Integrated Circuit Projects

Additionally, the 3162 controls the "Zero Set, and Gain through potentiometers R3 and R4. These allow you to zero the meter and adjust the voltage input range (9.99 volts, 99.9 volts, etc.). They tend to be a little sensitive, though, and more will be said about that in the next section. R1 also plays a role in voltage range selection (see parts list).

CIRCUIT DESCRIPTION

While a PCB will work for this device, the interface between the two chips is a little awkward, thus you might consider point-to-point wiring. Of course, if you plan to incorporate the design in a more elaborate scheme, then you will have to stay with the original wiring, but otherwise, the choice is there.

The size can be quite small, a couple of inches square, so this provides additional ease in plugging the circuit into new, or existing equipment. The recommended positive five volt power supply is usually available in most gear, but if not, an LM7805 regulator will quickly remedy that problem.

Price may be a concern, as the pair will set you back about ten dollars, where an ICL7107 can be purchased for under four. There are certain advantages to the RCA set, like a single positive voltage instead of the split supply required of many A/D converters. This makes for a far more convenient application.

Common anode displays are used, and they can be up to instrument size (0.56 inches). Usually, three separate LED units are multiplexed (all A's, B's, C's, etc. wired together) to form the display, but a surplus three digit common anode multiplexed display (whoa, try saying that quickly three times in a row) will serve nicely. Such displays can occasionally be found if you look hard.

As I stated earlier, the Zero, and Gain settings are a little on the precise or touchy side. This make the initial alignment a bit frustrating, but bear with it. A steady hand and patience will get you there. Fortunately, once set, the circuit is very stable, so you won't have to constantly re-adjust R3 and R4.

In **Figure 14-1**, I show the decimal point (D.P.) set at the second position, and this calibrates the display for the 99.9 volt range. However, since the gain control and R1 allow for voltage ranges between 0 and 100 volts, the decimal point can be set accordingly. Just be sure to drop the D.P. to ground through a 470 ohm resistor.

CONCLUSION

You will find that A/D converters are one of the most useful tools in the digital electronics arena. They allow you to translate virtually any form of analog information into data a digital system will understand, and through the use of drivers, display that information.

There are A/D converters in every size, shape and variety imaginable. Some have display drivers, some don't. Some produce 4 bit outputs, while others generate a full 8 bit byte, but they all share one thing in common. All transform analog signals to a digital format. That format can then be applied to the world of digital electronics; a very pragmatic and exciting world, indeed.

The RCA CA3162 is a classic example of this technology. It is small in size, reasonably priced, and highly reliable, all of which qualifies this chip as exceptionally functional. With its cousin, the CA3161, you can easily add a digital display to just about any project. Try it. You'll like it!

Integrated Circuit Projects

Project #15: 7107 Digital Thermometer

What You Will Need:

✔	U1	ICL7107CPL, 3 1/2 Digit LED Analog/Digital Converter
✔	DSPLY	4 Digit, Non-Multiplexed Common Anode LED Display

All resistors 5% carbon film.

✔	R1	470,000 Ohm, 1/4 Watt Resistor
✔	R2	1,000,000 Ohm, 1/4 Watt Resistor
✔	R3	25,000 Ohm Potentiometer
✔	R4	22,000 Ohm, 1/4 Watt Resistor
✔	R5	100,000 Ohm, 1/4 Watt Resistor
✔	R6	47,000 Ohm Potentiometer
✔	R7	470 Ohm, 1/4 Watt Resistor
✔	THM1	Negative Coefficient Bead Thermistor
✔	C1	0.22 Microfarad Mylar Capacitor
✔	C2	0.047 Microfarad Mylar Capacitor
✔	C3	0.01 Microfarad Mylar Capacitor
✔	C4	0.1 Microfarad Mylar Capacitor
✔	C5	100 Picofarad Ceramic Disk Capacitor

Integrated Circuit Projects

The 7107, designed by Intersil, has to be considered the Grand Daddy of all A/D converters, at least as far as we experimenters are concerned. Although the ICL7107CPL has been around for a while, and sometimes gets overlooked, it still remains one of the most versatile ICs ever conceived.

In this application, it is the basis for a remarkably accurate digital thermometer, but that is just the beginning for the 7107. Any quantity that is a voltage, or can be converted to a voltage, will serve as an input for the fundamental meter circuit, and the chip will then provide a digital representation of that input.

Best of all, the system is simple to construct, easy to use and employs a minimum of external components. The only real drawback is the split +/- 5 volt power supply necessary to run the meter, but with today's voltage regulators, or a couple of other options (zener diodes or the 7660 chip), this drawback is easily overcome.

CIRCUIT THEORY

First, let's look at the heart of this circuit, the ICL7107CPL. In a single 40 pin DIP there is a dual slope A/D converter, a reference, a clock, and seven segment display decoders and drivers. Additionally, the chip exhibits excellent anti-drift and error characteristics, and will give you a zero reading for a zero input.

The dual slope arrangement works something like this. Bear in mind this an abbreviated version, and not all there is to it. Anyway, basically a capacitor is charged with the input voltage (first slope), then discharged (second slope). Using a well regulated reference voltage as a starting point, the discharge is timed, and in this fashion, the clock pulses establish a digital representation of the input voltage.

Project #15: 7107 Digital Thermometer

Does that make sense? Well, as I said there is more to it, but it isn't essential information, so don't be too concerned if it eludes you. The primary point to remember is this process allows analog information to be introduced to digital systems, hence connecting them to the real world.

The reference voltage is generated through a zener diode and is controlled by the voltage divider R3/R4, see **Figure 15-1**. This needs to be set to one half the full scale reading, or in this case one volt, as the meter is configured for a two volt full scale.

The clock works with capacitor C5 and resistor R5 to set up the sample rate. With the recommended component values this will be three times a second, or a frequency of 48 kilohertz, but can be altered if the need arises. Additionally, the clock can be controlled by a crystal connected to pins 39 and 40, or an external oscillator connected to pin 40 and ground (pin 21).

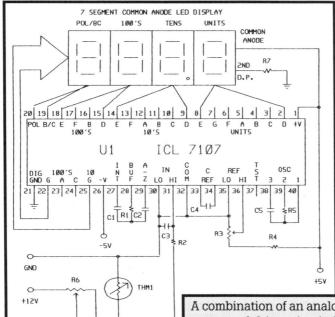

Figure 15-1. ICL7107 based digital thermometer.

A combination of an analog to digital converter and driver circuit, it can be used to measure any quantity that is a voltage or that can be converted to a voltage.

The decoder/driver section performs the same task as with the counter circuits; that is, to decode the BCD and drive the LED display. In this configuration, the display segments are addressed individually, thus a non-multiplexed display is employed. They can be up to instrument size (0.56 inches).

Our next consideration is the input transducer, or in this case, the temperature probe. Here again a voltage divider, comprised of potentiometer R6 and the thermistor THM1, is used. R6 calibrates the probe by establishing a reference, while the changing resistance of the thermistor indicates ambient temperature fluctuations.

Naturally, as THM1's resistance varies, so does the input voltage to the 7107 meter. Since we want the display reading to increase with a temperature rise, a negative coefficient style thermistor is needed, as its resistance goes down as the heat goes up.

As stated, one drawback to the 7107 is the necessity for a split power supply. Positive five and negative five volts are connected to pins 26 and 1 respectively, with pin 21 handling the ground. While this is an inconvenience, the performance of this chip well outweighs the extra components and effort.

Figure 15-2 illustrates both the main power supply and the separate temperature probe supply. Each can originate from the same transformer, but it is advisable to use individual secondaries, as you want to keep the 7107's input isolated, or floating with regards to its power source.

Both supplies utilize voltage regulators, with the positive 5 and 12 volts handled by 7800 series devices, and the negative 5 volts supplied by a 7900 series chip. In these schematics, the supplies are arranged as linear systems, with the transformer AC rectified by bridges, then filtered with large electrolytic capacitors.

Project #15: 7107 Digital Thermometer

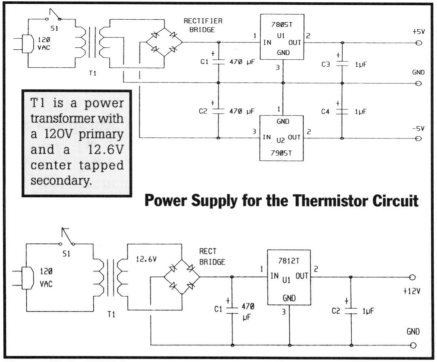

Figure 15-2. 7107 thermometer power supplies.

This filtered DC is now applied to the regulators which produce the stable output voltages. In the case of the split supply, the transformer's secondary winding center tap is used as the ground rail, and is not rectified.

The temperature probe supply is a standard dual rail configuration that provides the uniform output necessary for consistent and accurate readings. In both supplies, the smaller electrolytic capacitors across the outputs help filter out transient signals and prevent regulator oscillation.

CIRCUIT DESCRIPTION

These circuits are well suited to PCB construction, but as with most of the configurations in this book, they too can be point-to-point wired. That choice is pretty much up to you and your capabilities.

Integrated Circuit Projects

There is nothing particularly critical about any part of this project, just be careful to observe polarities, where applicable, orientation of the 7107, and proper construction technique. All regulators should be heat sinked to prevent overheating, and IC sockets are advisable for the 7107 and display.

The prototype has the second decimal point activated to provide 1/10 of a degree resolution. This is a matter of preference, and you can easily change that if desired. Be sure to connect the decimal point to ground through a 470 ohm resistor.

Once assembled, the circuit needs to be tested and calibrated. Apply power, and the display should be reading something except "00.0". If you are getting triple zeros this usually indicates the meter is not receiving an input, so check the temperature detector circuit for wiring errors and proper voltage.

If nothing at all is being displayed, then there is a problem with the meter itself. Again, the first place to look is the voltages. If that checks out, retrace your wiring, and watch for solder bridges with PCB construction. Also, try taking the 7107's test pin 37 high (+5 volts). If the chip is working correctly, and the display is properly connected, it will read "-188.8".

If all that fails, then regrettably, you may have a bad 7107. This is very rare, however, unless the chip has been exposed to excessive static charges, heat, or has been improperly wired. Normally, the 7107 is very durable if handled carefully.

With the meter working, the next step is to calibrate it for accurate temperature readings. First, adjust potentiom-

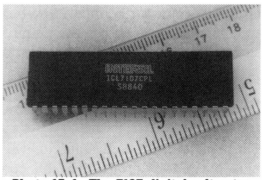

Photo 15-1. The 7107 digital voltmeter IC.

eter R3 to obtain a 1 volt reading between pins 35 and 36. This can be checked with a standard digital multimeter.

Next, prepare a glass of ice water, with both states (ice and water) present. No, not to drink! This will provide a very stable temperature reference of 32 degrees Fahrenheit, or 0 degrees Celsius. With the thermistor protected against moisture (sealed in epoxy or plastic, for example), place it in the ice water, and adjust potentiometer R6 until the display reads "32.0", or as close to that as you can get. If you want a Celsius reading, adjust for "00.0".

At this point, the "1/2" digit "1" is blanked, as it will not appear unless the temperature exceeds 99.9 degrees. Above that, the "1" does light, giving the unit the ability to read a maximum of "199.9" degrees.

Now the temperature meter is ready for action. The stability of the ICL7107 is such that no further adjustment should be necessary. That is, unless you change the circuit in some way (new thermistor, different power source, etc.).

CONCLUSION

I think the applications for a digital thermometer are more than obvious. If you do your own photographic processing and printing, need to monitor tank temperatures for tropical fish, work in chemistry, physics, or any number of other scientific fields, or just want to know the ambient room temperature, this device is for you.

There is no substitute for the clarity of these digital displays. With a second thermistor, and a single pole-double throw (SPDT) switch, you can locate the second sensor outside, and switch back and forth between inside and outside conditions. It won't tell you if it is raining out there, but you can't have everything.

Integrated Circuit Projects

Thus, this is a useful project. You could monitor the internal temperature of your computer, watch the Christmas tree lights for potentially dangerous overheating, keep an eye on the basement or attic for temperature problems, police the freezer for costly malfunctions, and the list goes on and on. Think about it. I feel certain you have a need for this gadget. If not, build it anyway, and send it to me (no COD's, please).

Project #16: 8038 Function Generator

What You Will Need:

- ✔ U1   8038CCPD, Waveform Generator/Voltage Controlled Oscillator

  All resistors 5% carbon film.
- ✔ R1   1,000 Ohm, 1/4 Watt Resistor
- ✔ R2   10,000 Ohm, Potentiometer (may be multi-turn)
- ✔ R3   100,000 Ohm, 1/4 Watt Resistor
- ✔ C1   0.001 Microfarad Ceramic Disk Capacitor
- ✔ C2   0.01 Microfarad Mylar Capacitor
- ✔ C3   0.1 Microfarad Mylar Capacitor
- ✔ C4   1 Microfarad Electrolytic Capacitor
- ✔ C5   47 Microfarad Electrolytic Capacitor
- ✔ S1   SP4T Rotary Switch

Integrated Circuit Projects

No test bench is complete without a function generator, and Intersil/Harris again comes to our rescue. On a single fourteen pin chip, they have assembled a device that will produce sine, square, and triangle waves, all simultaneously.

All right, this is a clever IC, and it is also very practical. As an experimenter, you know how often a stable frequency source is needed, and while you can breadboard a 555 timer for temporary service, what happens when you require a sine wave or triangle wave. That becomes a more complex problem.

Not so if you have this circuit handy. All three waveforms are readily available, and at frequencies from DC to 100 kilohertz. Each is clean except for the sine wave which displays a slight anomaly, usually not enough to cause any problems (barring perhaps the most stringent requirements).

CIRCUIT THEORY

The 8083 is a voltage controlled oscillator, and functions much like a timer, in that it charges and discharges a capacitor to provide the oscillation. A constant current source handles the charging, while a comparator detects a two thirds rise to full voltage and triggers a flip-flop. The flip-flop then begins the discharge cycle. At one third full voltage a second comparator re-triggers the flip-flop, and the charging starts again.

This process goes back and forth producing the square wave which is sent to an output buffer. The capacitor being charged and discharged by a constant current source results in a linear triangular waveform that also is buffered. Finally, the sine wave is the product of a converter stage that essentially rounds off the triangle peaks.

It is this rounding off action that results in an abnormal sine waves as the sides will tend to be straight rather than curved, but few applications are capable of detecting this. In much the same way, triangle and sine waves can be chopped at the top to successfully emulate square waves.

One of the really nice aspects of this configuration is that all three states, or waves, are present at the same time (different outputs, of course). Thus, any project or experiment that requires more than one waveform is easily accommodated by the 8038.

As for the outputs, when employing a 12 volt supply, the sine wave has a peak-to-peak level of one volt, the triangular form is four volts peak-to-peak, and the square wave will be two volts. Each provides more than sufficient signal for virtually any input requirement. Again, there will be exceptions, but hopefully you won't often encounter them.

The complete circuit can be assembled using a handful of external components, and the basic layout lends itself to modification and/or refinement, such as gain controls on the outputs. This will result in a more sophisticated piece of test equipment, but that is up to you. It is nice to have the options.

CIRCUIT DESCRIPTION

Overall size of this project can be quite small, although potentiometer R2 and switch S1 will have to be panel mounted, or at least accessible. As seen in **Figure 16-1**, S1 selects one of four capacitors that set the general range of the output. R2 then "fine tunes" the output to the exact frequency you need.

While I give the full range as 0 to 100 kilohertz, many hobbyists claim results of up to 1 megahertz. You might try using some small picofarad capacitors (100 picofarad, 10 picofarad, etc.) to increase the maximum frequency. However, as you increase the frequency, the wiring scheme will become more critical, due to stray inductance and capacitance.

Integrated Circuit Projects

This circuit will provide simultaneous square, triangular and sine waveforms at frequencies between DC and 100 kilohertz. The voltage can be between 5 and 15 volts, and all but the sine wave are clean, pure signals. The sine wave is close, but slightly flat on top. The outputs are 1VP-P for sine, 2VP-P for square and 4VP-P for triangle waves.

R2 can be a multiturn potentiometer to afford more precise frequency settings.

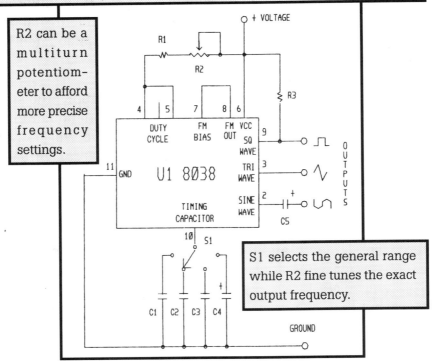

S1 selects the general range while R2 fine tunes the exact output frequency.

Figure 16-1. 8038CCPD based function generator.

The power source can be up to eighteen volts, so there is good latitude here. The circuit can be battery powered or operated off an AC regulated supply, whichever is more convenient. Power dissipation is rated at 750 milliwatts, rather high, but you probably won't be using the generator for very long periods. Hence, batteries should last for a respectable length of time.

A consideration here is that the output voltage will decrease if the supply voltage decreases. This is proportional in nature, thus for the full output, a twelve volt supply is recommended (six volts, for example, will provide only half the stated output).

One caution to be observed. The input voltage CANNOT exceed the supply voltage, or damage to the 8038 will occur. Take this one seriously, as you can fry this chip, and you don't want to do that.

CONCLUSION

So, pursue good technique, pay attention to detail, and you shouldn't have any trouble with this one. The end result will be well worth the meager effort involved in constructing this generator. That is not to diminish this project, as I always favor meager effort. The meagerer the better. In fact, simple IC projects are usually more reliable than their complicated counterparts.

If you have a frequency meter available, it will be of immense help in setting the output frequency. For that matter, take a second look at chapter five, as it has possibilities. Since you are working in primarily the audio range, what do you think? Could you modify that circuit for this purpose?

I think you can! Ah, go on. Give it a try! It's something to think about, anyway. A digital readout would be nice. OK, enough of the hard sell. Enjoy building the 8038 function generator, as I suspect you will not regret it. In addition to being able to make funny sounds, this device will serve you well on the work bench.

Integrated Circuit Projects

What You Will Need:

Basic Logic Probe:
- ✔ Q1   2N3904 NPN Silicon Signal Transistor
- ✔ LED1   Jumbo, T-1 3/4 Light Emitting Diode (any color)

   All resistors 5% carbon film.
- ✔ R1   10,000 Ohm, 1/4 Watt Resistor
- ✔ RX   470 Ohm, 1/4 Watt for 3 to 9 volt inputs
  1,000 Ohm, 1/4 Watt for 9 to 18 volt inputs

Inverted Based Hi/Lo Logic Probe:
- ✔ U1   74LS04, Hex Inverter
- ✔ LED1, 2   Jumbo T-1 3/4 Light Emitting Diodes (red and green)

   All resistors 5% carbon film.
- ✔ R1   330 Ohm, 1/4 Watt Resistor

Logic Probe with Hi and Lo Pulse Indication:
- ✔ U1   74LS04, Hex Inverter

Integrated Circuit Projects

- ✔ U2 — 74121, Monostable Multivibrator
- ✔ LED1 thru 3 — Jumbo, T-1 3/4 Light Emitting Diodes (vary the colors)

All resistors 5% carbon film.

- ✔ R1, 2, 4 — 220 Ohm, 1/4 Watt Resistors
- ✔ R3 — 330,000 Ohm, 1/4 Watt Resistor
- ✔ C1 — 0.01 Microfarad Mylar Capacitor

All right, let's explore the world of the logic probe, as it is an interesting world. Actually, the logic probe is a very simple device that performs a very important function. Simple in operation, but as we will see in this chapter, it comes in a variety of different configurations, all designed to do a special job.

If you plan to do much work in digital electronics, a logic probe, of one complexity or another, is a must have item. All too often it is necessary to know what the logic status (high or low) is at a given point within a circuit. Additionally, this device will allow you to trace the logic movement through a prototype, or piece of equipment under scrutiny.

Once constructed, the logic probe will be an invaluable tool when experimenting with all those nifty digital ICs and circuits.

CIRCUIT THEORY

In its simplest form, a logic probe can be two resistors, a transistor, and an LED. A transistor can function as a switch in which a potential at the emitter will pass to the collector whenever another potential is applied to the base.

Project #17: Logic Probe

In this arrangement, the base acts as a kind of gate that permits the flow of current from the emitter to the collector, but only if a positive voltage is present at the base. Notice in **Figure 17-1** that the positive voltage is applied to Q1's emitter through resistor Rx. An LED's anode is connected to Q1's collector, and the probe is hooked to Q1's base through a 10 kilohm resistor (I know this is getting a little redundant, but bear with me).

Now, using the probe to explore the test circuit, whenever you touch a point that is high (+ voltage), that contact activates Q1's gate, or base. This allows the positive voltage to flow from emitter to collector and on through the LED to ground, lighting the diode.

Thus, this circuit will let you know when the logic state is high. But what about a low state? **Figure 17-2**, shows a somewhat more complex circuit. In this one, either high or low will be indicated by separate LEDs.

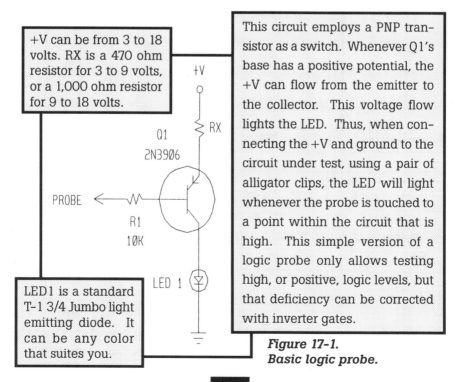

Figure 17-1.
Basic logic probe.

Integrated Circuit Projects

Here is how it works. When the probe encounters a high point, inverter U1A changes that to a low. Since the cathode of LED1 is connected to U1A's output, the circuit is complete, and LED1 lights.

When a low is applied to the probe, U1A changes it to a high. That will not light LED1, so it is introduced to the input of a second inverter, U1B, which changes it back to a low. U1B's output is connected to LED2's cathode, and when low, LED2 lights.

In **Figure 17-2**, one of the two LEDs will be on at all times, even when the probe is not touching a circuit point. This is because the probe will interpret no connection as a low, and light LED2. So, when testing with the probe, no change in the

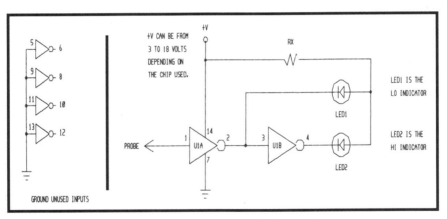

Figure 17-2. Inverter based Hi/Lo logic probe.

U1 can be either a 7404, 74LS04, or a 4049, depending on the type of circuit(s) to be tested. With standard 5 volt logic systems, the 7404 will work quite well. If, however, you plan to use the probe on higher voltage circuits, the 4049 is a better choice. RX will have to be changed as the voltage increases. For 3 to 6 volts, RX=330Ω, 60 to 9 volts, RX=470Ω, and for 9 to 15 volts, RX=1,000Ω. It is best to use separate color leads for the Hi/Lo indicators. Usually red for Hi and green for Low works well. Also, it is advisable to ground the inputs of the unused inverters as seen above. The pin numbers only apply to the 7404 or 74LS04 chips.

LED illumination means that you have touched a low point, as a high will cause the low indicator to extinguish, and the high LED to come on.

If the frequency is low enough (below 20 hertz), when you touch a point that is oscillating (pulsing), the two LEDs will alternately flash on and off. However, this is not a very reliable way to detect a pulse signal, as if the frequency is above 20 hertz, both LEDs will remain on constantly. At least, it will seem that way to your eyes (remember persistence of vision?).

This is where the circuit in **Figure 17-3** comes into play. Here, we add a multivibrator IC (U2) that will trigger an output when a pulsing signal is introduced to its inputs. Using the "not Q" output, which is low, LED3 will illuminate whenever the multivibrator is activated. You may also see some reaction in the other two LEDs, but the illumination of the third diode will positively indicate the presence of an oscillating signal.

Now you have the means to detect both high and low signals, as well as oscillations. Information of this nature can be invaluable when analyzing, and/or working with digital circuits.

CIRCUIT DESCRIPTION

Probably the best way to evaluate these three designs is to breadboard them and try them out. In all likelihood you will be most impressed with the probe in **Figure 17-3**, as the addition of the 74121 multivibrator enhances its value.

However, the simplicity of **Figure 17-1** allows for an admirable permanent addition to any microprocessor based project. By attaching an individual probe to each of the data and address bus lines, you can continuously monitor system activity.

Integrated Circuit Projects

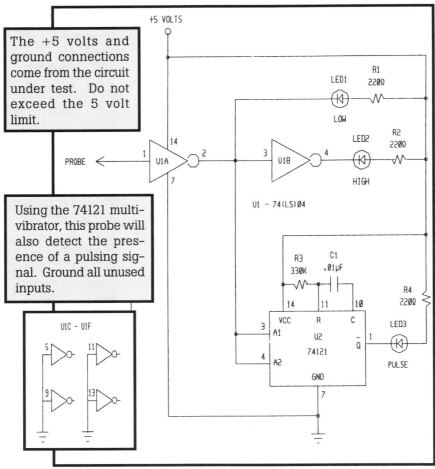

Figure 17-3. *Logic probe with Hi and Lo pulse indication.*

Of course, this will only indicate the high pulses. You could rig it so a second LED would light whenever the line went low, but that gets unnecessarily complicated. In reality, the first system is indicating a low by the LED not coming on.

For a quick and simple, but better featured approach, **Figure 17-2** fits the bill. If temporary service is all that is needed, this configuration can be breadboarded from a single 74LS04 hex inverter chip, and three discrete components. When done, the parts go back in their bins ready for another application.

Project #17: Logic Probe

One point worthy of mentioning here is that all unused inverter inputs should be tied to either ground or the positive rail. You will note in **Figure 17-2** and **Figure 17-3** that these inputs are grounded. This is done for power conservation and to prevent the circuit from absorbing noise. It's a good habit to adopt with all digital gates.

For the two later circuits, TTL logic ICs are used, in the form of the 74LS04 and 74121. As a reminder, TTL chips have the very narrow voltage tolerance of between 4.75 and 5.25 volts. Since the input voltage can not exceed the operating voltage, these probes can only test circuits with a 5 volt supply.

This is done by design, as most digital equipment falls within that parameter, at least as far as the logic section is concerned. However, if you need to check systems with a supply voltage higher than 5.25 volts, CMOS chips can be substituted for the TTL ICs.

Typically, the 4049 or 4069 hex inverters, and the 4538 dual multivibrator will do nicely, and being CMOS, they can handle supply voltages as high as 18 volts. Bear in mind, though, the input still must not exceed the supply.

Finally, in all cases power for the probe is borrowed from the circuit under test. Lengths of hook-up wire with alligator clips allow you to make the necessary power connections. Be sure, however, that you find the +5 volt supply if the probe is TTL based. Many digital circuits, computers for instance, have more than one supply voltage available.

CONCLUSION

So goes the tale of the logic probe. When working with digital circuits, few devices will prove more useful, and even fewer can be built so inexpensively. All in all, this is one piece of test equipment you will enjoy having around.

Integrated Circuit Projects

Modification, and/or special application is a snap with these designs, as was seen in the microprocessor bus line example. Put some thought to it, and you will come up with a double handful of similar ideas for employing a logic probe.

This is a very pragmatic endeavor. It goes without saying that the digital approach is a huge part of today's electronics, and there is no perceivable change in sight. If anything, it's going to get bigger. So, be prepared. It can be a lot of fun!

Project #18: Color Organ

What You Will Need:

✔	U1	LM386N-3, Low Voltage Audio Amp 500mw/9V
✔	U2 thru 4	LF351N, BIFET Operational Amplifiers
✔	D1 thru 3	1N914 Silicon Signal Diodes

All resistors 5% carbon film.

✔	R1, 5, 8	1,000 Ohm, 1/4 Watt Resistors
✔	R2	10,000 Ohm Potentiometer
✔	R3	10 Ohm, 1/4 Watt Resistor
✔	R4	4,700 Ohm, 1/4 Watt Resistor
✔	R6	500,000 Ohm Potentiometer
✔	R7	470 Ohm, 1/4 Watt Resistor
✔	R9	50,000 Ohm Potentiometer
✔	R10	47 Ohm, 1/4 Watt Resistor
✔	R11	10,000 Ohm, 1/4 Watt Resistor
✔	R12	5,000 Ohm Potentiometer
✔	C1	10 Microfarad Electrolytic Capacitor
✔	C2	0.05 Microfarad Mylar Capacitor
✔	C3	220 Microfarad Electrolytic Capacitor

Integrated Circuit Projects

- ✔ C4, 5   0.1 Microfarad Mylar Capacitors
- ✔ C6, 7   0.22 Microfarad Mylar Capacitors
- ✔ C8, 9   0.33 Microfarad Mylar Capacitors
- ✔ MIC   Electret Microphone
- ✔ RLY1 thru 3   SPDT Relays (see note)

Notes:

The relays will have to be of the same voltage as the operating voltage. Operating voltage can be from 5 to 15 volts. If high current loads are to be connected to the relays, it will be necessary to use the smaller on board relays to trigger heavy duty relays that can handle the loads.

This is a fun project that is easy to construct, and results in a great conversation piece. Once completed, this color organ will detect any sound near it, and activate one or more of three relays. Each relay represents low, middle or high audio frequencies, and can be used to control lamps, LEDs, lasers, strings of electric lights or anything else you can think of.

This circuit is most often used to add dancing lights to music systems, although the relays could manage most any electrical device. The term *color* refers to using lamps of various colors arranged in various displays. Depending on how you set up the displays, this arrangement can prove highly entertaining.

One feature that adds versatility to this system is that the circuit breaks the sound into three separate frequency ranges. If only high notes are playing, only that range is activated. The same is true of the low and middle ranges, so with a little imagination, the potential is limitless.

Project #18: Color Organ

The secret here lies in what is known as a bandpass filter. Three such filters are used with each one set for a different sound spectrum. Thus, the full range is covered in three very useful pieces.

CIRCUIT THEORY

Filters are a science all of their own. They are utilized in an enormous number of electronic applications to filter everything from the 60 hertz hum of standard AC power lines to transient signals and electromagnetic fields. While simple (for the most part) and less glamorous than many other areas of electronics, they play a vital role in keeping things working.

They have many names, Sallen-Key, high pass, notch, bandpass and so forth, but it is this last category that permits the project to function. In our case, as seen in **Figure 18-1**, the bandpass filter is a simple device using resistors and capacitors to set parameters, and an op-amp to create a window, or range, of passable frequencies.

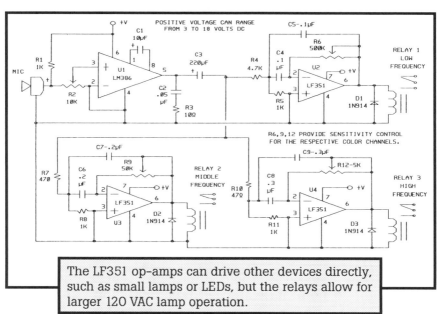

The LF351 op-amps can drive other devices directly, such as small lamps or LEDs, but the relays allow for larger 120 VAC lamp operation.

Figure 18-1. Three channel color organ.

This window concept is not exclusive to filters. It is employed in frequency counters, comparator circuits, voltage control and monitoring, as well as other applications. Remember chapter five and the temperature detector? In that configuration the op-amp creates a window of acceptable temperatures.

For the color organ we calibrate each of the three filters for one of the separate ranges using tuning networks C4/R5, C6/RB, and C8/Rll. In this fashion, the filters will only respond to frequencies within their windows, and then activate their relays.

The second part of this circuit utilizes our friend the LM386 audio amplifier. This chip amplifies the sound picked up by the microphone to a level the bandpass filters can work with. A quick glance tells you that this is the standard 386 layout, with potentiometer R2 acting as the volume control.

The output from a stereo or other amplified device could be sent directly to the filter input, but an audio transformer would be needed. Since the stereo's output is in the eight ohm range, you would want to hook the transformer's eight ohm winding there. The secondary winding, usually one to two thousand ohms, would be connected to the filter input, and ground.

The microphone shown in **Figure 18-1** is an electret variety, but other types could be employed if a pre-amp stage is added. An additional point to mention is that the three signal diodes (D1-3), in parallel with the relay coils, serve an important purpose. When power is removed from a coil, its collapsing magnetic field can generate dangerous voltage spikes. These diodes protect the ICs by absorbing those spikes.

CIRCUIT DESCRIPTION

This is a more complex circuit than some, but it still can be wired point-to-point if desired. PCB construction will provide additional durability, however you may find the other method more convenient.

Project #18: Color Organ

Either way, it is a standard layout. Keep the technique clean, but there is nothing critical about the configuration. It's not like we were dealing with microwaves here.

Probably your biggest concern will be the relays. First, their coils will have to be compatible with the supply voltage. For example, if you are using 12 volts, the highest the coil rating can be is 12 volts, but that may give you some trouble.

Since the op-amps probably will not have outputs quite as high as the supply voltage, some relays may not want to activate. Thus, it would be wise to employ 9 or 10 volt relays instead of 12 volt. In most cases the coils will be able to handle the extra voltage. However, if the relays become hot, use series dropping resistors (typically 1 kilohm, or thereabout.)

The next consideration will be the load wattage. The relay contacts will be rated for a certain amperage, which when multiplied by the voltage will give you the wattage. Let's say you want to light a 200 watt bulb from a 120 volt AC line. Here you have the voltage and wattage, so you need to determine the amperage.

If wattage is equal to amperage multiplied by voltage, then amperage is equal to wattage divided by voltage. So in this case you will need to divide 200 by 120 to get 1.7 amps. This means that the relay contacts have to be rated at a minimum of 1.7 amps, but for safety, bump that up to 2 or 3 amps.

The same approach will apply to whatever load you decide to use. Merely divide the load's wattage by the voltage in use, and you'll have the amperage rating for the relay. On-board relays will accommodate loads up to about 5 amps, but what if you need more current than that?

Relays are available that will handle 15, 20, even 30 or more amps, but the rub is the coils. While they may be rated at 12 volts, they often require a higher amperage than the op-amps can provide. Thus, you will need to use the on-board relays to activate a second set of heavier duty relays.

Employing an additional power supply, the first contacts apply this higher amperage potential to the second relay's coil. This, in turn, activates the second set of contacts, and your higher amperage loads. Using such a relay daisy-chain scheme, the total load can be quite high in amperage.

To furnish even more versatility, potentiometers R6, R9 and R12 are included to control the sensitivity of their respective channels. Having these available allows for some interesting variations in the overall display. For example, you could adjust the circuit to provide very bright lighting for the low tones, and low illumination for the medium tones, or low lighting for the high tones, or...well, you get the idea.

CONCLUSION

All right, don't even try to tell me you haven't already thought of some applications for this one. The possibilities are endless. Think what it will do for your den, or the Christmas tree, or the kids' room, or Halloween. You see what I mean.

In all honesty, this circuit can produce anything from a very demure indicator for stereo equipment to a full fledge carnival, or disco atmosphere. It just depends on how far you want to take it.

Of additional value here is another lesson in the adaptability of op-amps. Like the LM555 timer, they can perform numerous tasks, and are well worth understanding.

Project #19: Laser Diode Driver Circuits

What You Will Need:

- ✔ Q1 — 2N3906 PNP Silicon Transistor
- ✔ Q2 — 2N3904 NPN Silicon Transistor
- ✔ LDM1 — Visible Laser Diode Module
- ✔ U1 — LM317T Adjustable Voltage Regulator
- ✔ LED — T-1 3/4 Jumbo Light Emitting Diode (any color)

All resistors 5% carbon film.

- ✔ R1 — 1,000,000 Ohm Potentiometer
- ✔ R2 — 100 Ohm, 1/4 Watt Resistor
- ✔ R3 — 5,000 Ohm Potentiometer
- ✔ R4 — 220 Ohm, 1/4 Watt Resistor
- ✔ R5 — 470 Ohm, 1/4 Watt Resistor
- ✔ C1, 3 — 1 Microfarad Electrolytic Capacitors
- ✔ C2 — 10 Microfarad Electrolytic Capacitor

Integrated Circuit Projects

When the twentieth century comes to an end, one device will dominate the list of inventions, the laser. If not considered the most significant development of at least the last half of the century, it will surely rank in the top three. Few other inventions have so transformed and benefited life as we know it, and for this reason, lasers will take their rightful place in history.

In the way of a short history, the first working laser (Light Amplification by Stimulated Emission of Radiation) was developed by the Bell Laboratories as a result of Albert Einstein's 1913 Stimulated Emission of Radiation theory. This device used a rod of ruby colored synthetic spinel, and was activated by bright flashes of light from an energized xenon tube.

The result was a pulse of light, highly collimated and coherent, that was capable of burning holes through many tough materials (stone, glass, metal, etc.) With this instrument functional, the world of lasers and laser light was born.

Today's lasers come in a multitude of sizes, shapes, types, and strengths, with most of them being used for purposes other than drilling plate steel. The gaseous, or ion, variety is probably most familiar to the average person, as it is hard to buy anything these days without it passing over a Universal Product Code (UPC) scanner (those windows in the top of the checkout counter that make a beep when the clerk drags your item over it).

If you look carefully, down inside the contraption, you will see a pattern of crisscross threads of red light, and these are produced by a Helium-Neon gas laser. A similar light can be seen emitting from the end of the pistol shaped UPC hand units, and this is the product of a different type of laser; the laser diode (a solid state laser).

It is this variety that we are concerned with. Laser diodes have emerged over that last fifteen years or so, and have become

Project #19: Laser Diode Driver Circuits

widely used in all kinds of stuff. Aside from the UPC scanners, they are found in video and music disk players, CD-ROM drives on computers, an assortment of specialized optical reading equipment, and of course, the laser pointer!

Unlike the Helium-Neon lasers, the diodes do not require high voltages (2 to 5 kilovolt nominal) to function. They usually work in the 1.75 to 2.5 volt range. However, they cannot be connected directly to a battery. The reason for this will become clear shortly.

What they do require is a driver circuit. These can be fashioned from op-amps, but transistors are often a better bet as they provide higher amperage ratings. Many of the diode lasers want to see 40 milliamps or better to reach their threshold, or firing point. Thus, this circuit is designed with transistors in an effort to make it more universal.

The reason for the electronics involves the fact that diode lasers are highly susceptible to power surges, transients, and static charges. In order to avoid damaging or even destroying the delicate laser diode, it is necessary to control the current flow through it. This is accomplished by integrating a photodiode (PD) within the laser diode (LD) substructure. **Figure 19-1** illustrates this in the boxed-in section of the schematic (at bottom), which signifies a laser diode module.

The photodiode detects changes in the brightness of the laser diode, and adjusts the current accordingly. This not only protects the laser, but also keeps the output consistent. So, let's take a closer look at just how the circuit achieves this task.

CIRCUIT THEORY

Referring to **Figure 19-1**, you will note that the anode of the photodiode is connected to the base of transistor Q1. The cath-

Integrated Circuit Projects

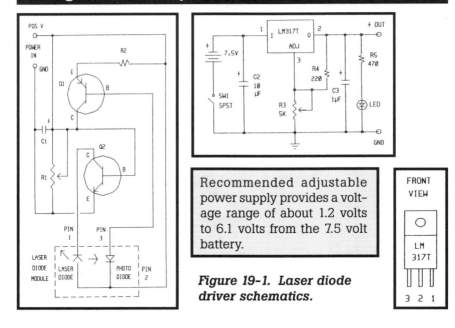

Figure 19-1. Laser diode driver schematics.

Recommended adjustable power supply provides a voltage range of about 1.2 volts to 6.1 volts from the 7.5 volt battery.

ode, in turn, is hooked to the positive voltage source, and this configures the photodiode in a reverse bias mode. As the light on the diode increases, the current flow decreases.

This change in current determines the amount of flow from Q1's emitter to collector. If you remember, the transistor acts as a gate controlled by the base. The collector is then run to the base of Q2, which controls the current flow to the laser diode's cathode. With the anode connected to the positive rail, the laser is forward biased, and as with other LEDs, glows with an intensity proportional to the applied current.

In short, the photodiode watches the laser's output, and if it gets too bright, the PD turns it down. If the laser becomes dim, the photodiode turns up the juice. In the end, just the right amount of current is allowed to flow through the laser, thus protecting it from damage and keeping the light amplitude uniform.

Potentiometer R1 functions as a maximum current adjustment, but for most conventional visible light laser diodes, it can be

Project #19: Laser Diode Driver Circuits

set in the neighborhood of 300,000 ohms. That usually places the current where it needs to be. Some of the older diodes might require higher current values to achieve full power, and this can be controlled by Rl.

Capacitor C1 acts as additional transient protection. As stated earlier, these devices can easily be damaged by transitory energy on the power line, so this component is a prudent circuit addition.

CIRCUIT DESCRIPTION

A quick look at **Figure 19-1** reveals a very simple circuit. with two transistors, two resistors, and a capacitor, the configuration is clean, and lends itself well to both point-to-point wiring and PCB construction. Either one or breadboarding works. The choice is yours.

Figure 19-2 provides a pinout diagram for most visible light laser diodes currently available. I say most as there may well be other configurations out there, but in my experience, this is the layout encountered. The orientation notch marks the common, or number 2 pin, although some packages will have a notch at the 1 and 3 pins as well. Thus, it is best to identify the pins from their location rather than from the notch.

On a quick note, in this instance, when I refer to visible light I am talking about somewhere in the red light spectrum (780 to

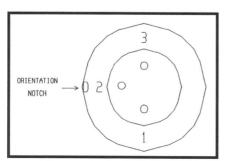

Figure 19-2. Standard visible light laser diode pinout.

630 manometers). Many of the laser diodes operate in the infrared range (below 780 manometers) which is not visible to the human eye.

Laser diodes come in two sizes, the more common 9 millimeter packages and the smaller 5.6 millimeter cases. Functionally, there is no difference between the two. The small package was developed primarily for laser pointers and other applications where compact size is an advantage. When you order, you usually have to take pot luck.

Again, let me emphasize that these modules are extremely sensitive to static electricity. Use all appropriate anti-static procedures (grounding yourself and work area, conductive foam, etc.) when handling and storing laser diodes, or you will pay a nasty price (Kentucky fried diode). Once installed in the driver, the risk is substantially less, but still use caution.

Now that the driver circuit is out of the way, let's direct our attention to the power supply. Unless you need to run the system for extended lengths of time, battery operation is highly recommended. I suggest batteries because they rarely, if ever, generate spikes or surges (remember, we don't want those), and even though the lasers draw between 30 and 100 milliamps, the battery pack will provide more than acceptable service life.

Photo 19-1. Top and left: common 9mm size. Right: newer, compact 5.6mm design.

The power source is applied to the input of a generic LM317T adjustable linear configuration, which acts to regulate the output voltage. From 7.5 volts, this supply will deliver 1.2 to about 6.1 volts, and that should cover the requirements of virtually all laser diodes.

Project #19: Laser Diode Driver Circuits

If you have to operate the system from wall current, make very certain the power supply is free of transient signals and is protected against power surges. I know you're getting tired of hearing this, but I really can't stress it enough. If these precautions are not taken, the result will be woeful.

With both sections constructed, hook them together, and you have your completed driver assembly. They can be incorporated on the same board if desired, however the variable power supply will come in handy with other projects, so I have built them separately.

Now is the time to install the laser diode. I have personally found it best to use SIPs, arranged as a socket, but the diode can be soldered directly to the board. If you go that route, use a grounded tip soldering iron, and as little heat as possible. Here again, all the gremlins need to be warded off (static, line surges, overheating the semiconductor junction, and so forth).

When connecting the laser, by whatever means, be sure to observe pin positioning. Reversing the 1 and 3 pins probably will not damage the module, but it won't work. With a socket arrangement, though, this is hard to do, or easily corrected if you should manage it.

Photo 19-2. Completed laser driver circuit.

Once everything is good to go, set potentiometer R3 to the lowest voltage output (1.2 volts) and turn on the power. Now, gradually advance R3 until you begin to see a slight glow inside the laser. CAUTION: DO NOT LOOK DIRECTLY INTO THE WINDOW OF THE LASER MODULE! View the activity from as steep an angle as possible, as these diodes put out between 3 and 6 milliwatts of power and are a hazard to the eye. Remember, these are Gallium/ Arsenide junctions, with a little Aluminum thrown in for fun, and they do emit a certain amount of infrared energy.

This first glow is known as threshold, and is equivalent to a Helium-Neon tube beginning to lase. Carefully continue to advance the voltage until the diode is glowing brightly. At this point, it is wise to check the power supply output with a voltmeter. I have found that most laser diodes can handle up to 6 volts as long as the current doesn't get too high, but you do not want to burn it up.

Whenever ordering a laser diode, it is best to request the data sheet, even if it costs a buck or so. A lot of valuable information is contained therein that can help guide you in using the laser. There will usually be ratings for threshold and operating voltages and currents. As this varies greatly from one diode to the next, the technical data is well worth the investment.

CONCLUSION

If all has gone according to the plan, you now have a working diode (solid state) laser. Since the divergence (beam spread) of these devices is far greater than other lasers, don't expect to see the small dot on the wall you saw when someone showed you a Helium-Neon system in action. It will take collimating optics (an old camera lens or magnifying glass) to accomplish that.

However, you will see the telltale signs of true laser light; that is, coherency (single color) and speckling (a glittering effect due to the light reflecting off surfaces of different depth). It is these qualities and others that make laser light so special.

Let me again remind you of the possible eye hazard lasers can present. The properties of coherency and collimation produce a very concentrated beam of light that, in conjunction with the focusing ability of your eye, can and will burn the retina if viewed directly. This can also result in permanent damage to your vision. As long as proper precautions are observed, lasers are safe and pragmatic instruments of great educational value.

Project #20: Audio Tone Generator

What You Will Need:

✔	U1, 2	LM555 Timer/Oscillator

All resistors 5% carbon film.

✔	R1	1,000 Ohm Potentiometer
✔	R2, 5	100,000 Ohm, 1/4 Watt Resistors
✔	R3	10,000 Ohm Potentiometer
✔	R4	4,700 Ohm, 1/4 Watt Resistor
✔	R6	100 Ohm, 1/4 Watt Resistor
✔	C1, 8	1 Microfarad Electrolytic Capacitors
✔	C2	2.2 Microfarad Electrolytic Capacitor
✔	C3	4.7 Microfarad Electrolytic Capacitor
✔	C4	10 Microfarad Electrolytic Capacitor
✔	C5	0.01 Microfarad Mylar Capacitor
✔	C6	0.04 Microfarad Mylar Capacitor
✔	C7	0.1 Microfarad Mylar Capacitor
✔	SPKR	8 Ohm Speaker
✔	S1, 3	SP4T Rotary Switches
✔	S2	SP3T Rotary Switch

Integrated Circuit Projects

The LM555 will be used again in this chapter. This time, however, we will use it to produce some interesting sound effects. In addition to a continuous tone, this configuration will produce tone bursts and dual alternating high-low tone patterns.

This makes it great for games, toys, video and audio productions, special sound effects, as well as a variety of other applications. The circuit is easy to construct, using off-the-shelf components, and can be kept small for those tight places.

CIRCUIT THEORY

Take a look at **Figure 20-1**, and you will see that this circuit is comprised of two LM555 timer/oscillators (U1 and U2). Both are arranged in the astable mode, with U2's reset controlled by U1's output. U2's output is connected to an 8 ohm speaker to provide the audio.

In order to understand how this circuit works, let's examine some of the 555 characteristics. For the astable configuration,

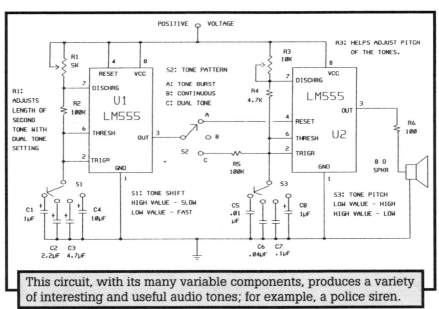

This circuit, with its many variable components, produces a variety of interesting and useful audio tones; for example, a police siren.

Figure 20-1. LM555 based audio tone generator.

Project #20: Audio Tone Generator

reset pin 4 is usually kept high, but taking pin 4 low will reset the timing cycle. So, by connecting the first oscillator's output to the second chip's pin 4, each time that output goes low the second oscillator will reset.

Since the astable output is low during the timing period, oscillator two is reset, and its cycle determines the output tone. This periodic resetting of U2 results in a tone burst effect.

If pin 4 is left floating (not connected), then the astable configuration will continually retrigger itself, and produce a constant tone. This is accomplished by connecting the threshold (pin 6) to the trigger (pin 2). In this mode, the first oscillator has no influence over the second.

For dual tone operation, U1 sets up both the duration of the second tone and the speed at which the tones shift. Potentiometer R1 controls second tone interval, while the timing capacitors (C1-C4) handle the shift. A slow change emanates from a large value capacitor, and the reverse with a small capacitor.

In all cases, the timing capacitors (C5-C8) and charge resistor (potentiometer R3) control the tone pitch. The larger the capacitor, the lower the tone, with the same being true for the setting of R3.

In sum, the first oscillator only plays a role in the operation of the tone burst and dual tone modes, not the continuous tone mode. The interaction is by way of directing the first oscillator's output to either the charge/discharge network or reset pin of timer two. With that output left open, the second oscillator operates independently of the first.

All of this isn't really hard, it just sounds that way. Basically, you have a circuit similar to an intervalometer, in which one oscillator triggers another. Instead of tripping relays, these oscillators control the duration and pitch of audio tones.

CIRCUIT DESCRIPTION

As **Figure 20-1** illustrates, this device is not complicated. Either wiring scheme can be used with good results, and the layout is hardly critical. As always, sound construction technique is encouraged.

If desired, a LM556 dual timer chip can be used in place of the two LM555 ICs, but clarity of circuit operation may suffer slightly. All other components are run of the mill, which makes things easier.

In the event you are planning on a specific sound, such as a police siren, it is best to breadboard the design, and determine the capacitor values. That will eliminate the need for the rotary switches. The potentiometers can be small PCB mounted types, and all of this will keep the overall size to a minimum.

For a more versatile arrangement, the circuit can be enclosed in a project case with the potentiometers and switches panel mounted. This will provide full access to all controls, and result in a very handy work bench tone generator.

As for power, the circuit will operate for limited periods of time off a standard 9 volt alkaline battery. However, if you plan on using it for extended service, an AC based power supply is recommended. The old stand-by linear regulated configuration will perform admirably. Potentials from 6 to 15 volts DC are acceptable, but the chips will run cooler at a lower voltage.

The final output can go directly to an 8 ohm speaker, to a jack, or to a switch that will send it in either direction. Again, for a single sound dedicated system, a small speaker is probably best. For the work bench version, something more elaborate is most likely in order. You may very well want both a speaker, and a jack connected to U2's output.

Project #20: Audio Tone Generator

Now that you have this gem built, let's play with it. With the power on, set switch S2 to the continuous tone mode. Select one of the timing capacitors with switch S3, and you should be hearing a tone. Switch back and forth through S3's settings, and you will notice the change in pitch.

Next, rotate potentiometer R3, and again, you will hear the tone pitch change. That little exercise illustrates the basic performance of the generator. Next, we want to bring in a second function.

Leave S3 and R3 at their last settings, and change S2 to the tone burst mode. Now you will hear a repeating tone separated by silence. Changing switch S1 will change the speed at which the tone repeats.

Last, let's switch to the dual tone mode. This will bring an alternating high-low tone that repeats over and over again. Adjust R1 and the duration of the second tone will change. Rotating switch S1 causes the speed of the tone shift to fluctuate, and of course, R3 and S3 will allow you to control the pitch.

This setting is the most entertaining, as a wide variety of tone combinations can be achieved. These will range from the familiar police sire to some really bizarre sounds. In this mode, the possibilities for games, toys and special effects are unbelievable.

CONCLUSION

All right, there you have it all you sound freaks. Just wait 'till that next video production, you'll tear them up, eh? Or, Halloween, yeah, what about Halloween? You dog you.

Now, you all can't say I don't take care of you. I mean, there's something in this book for everyone. Well anyway, all kidding

Integrated Circuit Projects

aside, this is a handy gadget that's a sure fire attention-getter. That in itself opens up opportunities along the lines of advertising a product or service.

So, break out the junk box, and whip this one up. It's easy to build, and a treat to play with. How about a robot that sounds like a fire engine? Hey, that ought to keep the kids or the cat quiet, at least for a little awhile.

Project #21: TV Modulator

What You Will Need:

✔	U1	MC1374P, TV Modulator (JDR Micro)
✔	D1	1N4001 Rectifier Diode
✔	D2	1N914 Signal Diode
✔	L1	4 Turns, #22 Wire on 1/4 Inch Form
✔	L2	40 Turns, #36 Wire on 3/16 Inch Form
✔	L3, 4	0.22 Microhenry Fixed Inductors

All resistors 5% carbon film.

✔	R1	10,000 Ohm, 1/4 Watt Resistor
✔	R2 thru 4	470 Ohm, 1/4 Watt Resistors
✔	R5	6,800 Ohm, 1/4 Watt Resistor
✔	R6	33,00 Ohm, 1/4 Watt Resistor
✔	R7, 11	2,200 Ohm, 1/4 Watt Resistors
✔	R8	75 Ohm, 1/4 Watt Resistor
✔	R9	56,000 Ohm, 1/4 Watt Resistor
✔	R10	220 Ohm, 1/4 Watt Resistor
✔	R12	560 Ohm, 1/4 Watt Resistor
✔	R13	180,000 Ohm, 1/4 Watt Resistor
✔	R14	30,000 Ohm, 1/4 Watt Resistor

Integrated Circuit Projects

- ✔ C1, 3, 8, 9, 13 — 0.001 Microfarad Ceramic Disk Capacitors
- ✔ C2 — 5 to 25 Picofarad Variable (trimmer) Capacitor
- ✔ C4 — 56 Picofarad Ceramic Disk Capacitor
- ✔ C5 — 0.01 Microfarad Ceramic Disk Capacitor
- ✔ C6 — 50 Picofarad Ceramic Disk Capacitor
- ✔ C7 — 120 Picofarad Ceramic Disk Capacitor
- ✔ C10, 12 — 22 Picofarad Ceramic Disk Capacitors
- ✔ C11, 14 — 47 Picofarad Ceramic Disk Capacitors
- ✔ C15 — 10 Microfarad Electrolytic Capacitor
- ✔ C16 — 1 Microfarad Electrolytic Capacitor

Does "News at Eleven" give you any hint of what's to come in this chapter? Perhaps not, but it should evoke a metal picture of a certain pastime we all know and love (well, maybe), and that should put you in the ball park.

Television! Yes, television, and this project will allow you to introduce your own programming to your own TV set (closed circuit, of course). Now you can hook up a television camera or camcorder to the ol' tube, and have your very own "News at Eleven". For that matter, any other show that grabs you.

I know. You're saying, "Heck, I can buy one of those things on the surplus market for five dollars." This is true, but think of the satisfaction, pride and self-esteem you will experience when you construct this device with your own two hands!

This is a very educational project that results in a very useful device. Besides, video circuits are always fun, and easily modified, and you can't say that about your store bought box.

Project #21: TV Modulator

CIRCUIT THEORY

As **Figure 21-1** illustrates, the heart of this project is the Motorola MC1374P TV modulator. Adding the various discrete components turns this chip into a micropower television transmitter that can be tuned for either VHF channel 3 or 4.

This should sound familiar, as most video games, the early Commodore computers and many other video related devices, use these same two channels for the interface. It is a highly practical approach to employ conventional home televisions as monitors.

In our design, this is accomplished by connecting a resistor, capacitor, and diode network to either ground for channel 3, or the positive rail for channel 4. Either way, you will deliver both video and audio to the television set.

The MC1374 provides a stable and efficient method of producing a TV compatable signal. Either channel 3 or 4 can be selected by switch S1. Standard video and audio inputs are combined to produce the proper output. Try using a short length of wire as the output for a short range wireless transmitter.

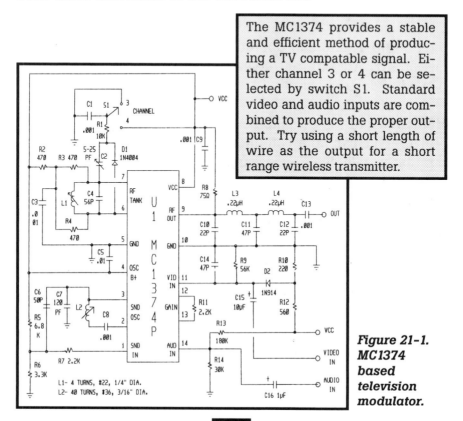

Figure 21-1. MC1374 based television modulator.

Integrated Circuit Projects

Rather than try to explain what each components does in this circuit (I'm not sure I know myself), let's just say that the configuration is Motorola's recommended layout for this chip. If you have any experience with high frequency RF circuits, you will notice a similarity, especially with single chip designs.

As would be expected, the various resistors, capacitors, diodes, and coils are support components, with the variable capacitor and variable coils used to fine tune the circuit. Actually, the stability of the MC1374 is such that very little adjustment is required.

For some specifics about the chip, it is a 14 pin DIP which contains an RF oscillator, RF modulator, and a phase-shift FM modulator. Power comes from a single supply that can range from 5 to 12 volts, and for best results, a regulated supply is advised.

The RF oscillator is controlled by a tank circuit connected to pins 6 and 7, and can operate at frequencies as high as 105 megahertz. An external crystal oscillator could also be employed, if desired.

Modulator gain is accomplished by a resistor across pins 12 and 13. This can be a potentiometer, for variable gain control, but that usually isn't necessary.

The MC1374 displays wide dynamic range and low audio distortion, which makes it ideal for our application. Additionally, the chip can function as an FM test signal source, or could be employed in cordless telephone base stations.

Basically, that is all that needs to be said about this IC. For the purists among us, the data sheet does go into more detail about the 1374's operation, but it's engineer's stuff (you know, graphs and charts and formulas).

Project #21: TV Modulator

CIRCUIT DESCRIPTION

As **Figure 21-1** reveals, there are a number of external components to this circuit. With fourteen resistors, sixteen capacitors, four coils, and two diodes, the total count is higher than many chips, but that is not unusual for this type of device.

PBC construction is recommended, as the higher operating frequencies make point-to-point wiring problematic. All that copper wire running all over the place creates too much transient capacitance and inductance.

Coils L1 and L2 have to be hand wired on tunable forms. However, this isn't difficult, with L1 consisting of four turns of number 22 wire on a 1/4 inch form, and L2 using forty turns of number 36 wire on a 3/16 inch core. We are talking about enameled wire here, not bare bus wire.

With that done, the next step is to assemble the circuit. All components are installed in the normal fashion, and it is best to employ an IC socket with the MC1374. See **Figure 21-2** for pinouts.

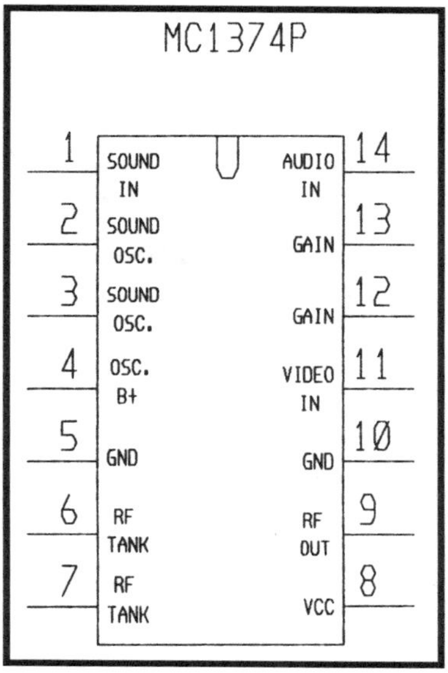

```
            MC1374P
        ┌─────────────┐
    1 ──┤ SOUND  AUDIO├── 14
        │ IN      IN  │
    2 ──┤ SOUND       ├── 13
        │ OSC.   GAIN │
    3 ──┤ SOUND       ├── 12
        │ OSC.   GAIN │
    4 ──┤ OSC.  VIDEO ├── 11
        │ B+      IN  │
    5 ──┤ GND     GND ├── 10
        │             │
    6 ──┤ RF      RF  ├── 9
        │ TANK   OUT  │
    7 ──┤ RF          ├── 8
        │ TANK   VCC  │
        └─────────────┘
```

PIN 1	Sound Carrier In
PIN 2	Sound Carrier Oscillator
PIN 3	Sound Carrier Oscillator
PIN 4	Carrier Oscillator B+
PIN 6	Radio Frequency Tank
PIN 7	Radio Frequency Tank
PIN 8	VCC (positive voltage)
PIN 9	Radio Frequency Out
PIN 10	Ground
PIN 11	Video In
PIN 12	Modulator Gain
PIN 13	Modulator Gain
PIN 14	Audio In

Figure 21-2. TV modulator MC 1374P pinout diagram.

Now you will want to align the RF tank and sound oscillator circuits, and this is best done as follows. Connect the power and video/audio sources, set switch S1 to whichever channel (3 or 4) you want to receive the signal, and hook the output (pin 9) to the television's antenna input.

First adjust coil L1 for the clearest picture, then set L2 for the best audio quality. Variable capacitor C2 might require some adjustment as well, so it is best to set it at mid scale, and if necessary, work from there.

That is all there is to it. Now you can introduce any standard video and audio source to pins 11 and 14, respectively, and said programming will appear on the television monitor. Included here could be area surveillance, video games, or your very own TV productions.

On a final note, try connecting a short whip, or length of wire (6 to 8 inches) at the modulator's output (pin 9). This will act as an antenna for wireless transmission of the video/audio input. I won't spoil this experiment by giving you anticipated results, but I think you will be pleased by the device's performance.

CONCLUSION

This is an extremely useful tool when experimenting with video projects. Most television cameras produce a composite baseband output which has a peak-to-peak value of roughly 1 volt. This is a video standard, quite acceptable with video monitors, but not directly compatible with television receivers.

The same is true of many other video sources, so unless you have a video composite monitor, you are unable to view the signals. However, by now it should be obvious this unit will rid you of that obstacle. Through the magic of the MC1374P, all signals are yours to see. This single device makes setting up a video experimental lab a breeze.

Project #22: Touch-Tone Generator

What You Will Need:

Tone Generator:

- ✔ U1 — TCM5089N, Integrated Tone Dialer
- ✔ U2 — LM386N-3, Low Voltage Audio Amplifier 500mw/9V

All resistors 5% carbon film.

- ✔ XTAL — 3.579545 Megahertz Color Burst Crystal
- ✔ R1 — 10,000 Ohm Potentiometer
- ✔ R2 — 1,000 Ohm, 1/4 Watt Resistor
- ✔ R3 — 10 Ohm 1/4 Watt Resistor
- ✔ C1 — 220 Microfarad Electrolytic Capacitor
- ✔ C2 — 0.05 Microfarad Mylar Capacitor
- ✔ SPKR — 8 Ohm Speaker (see text)
- ✔ KEYBRD — 12 Key Stardard X/Y Matrix Telephone Keypad

Telephone Line Interface:

- ✔ MOV1 — Metal Oxide Varistor (Jameco V130LA2)
- ✔ BR1 — W02G Full Wave Bridge Rectifier or Equiv.
- ✔ D1 — 1N4735, 6.2 Volt Zener Diode
- ✔ R1 — 22 Ohm, 2 Watt Resistor (5% carbon film)

Integrated Circuit Projects

With the invention of dual tone multi-frequency (DTMF) telephonery, our way of life has changed dramatically. Not only can you play tunes with the telephone, but it is much faster and easier to use. Furthermore, any number of sadistic companies can play games with us. You know, "If you have a touch tone phone press 1." This usually results in another message telling you to push more buttons, or you get disconnected. Come to think of it, maybe we ought to forget this project, and go back to the rotary phones.

This project will produce a very interesting and educational device that can help you better understand modern telephones. Additionally, we will explore some other areas where DTMF simplifies things, so the chapter won't be a complete loss.

CIRCUIT THEORY

Figure 22-1 illustrates the schematic diagram of this DTMF tone generator. As with all touch-tone systems, the trick is to

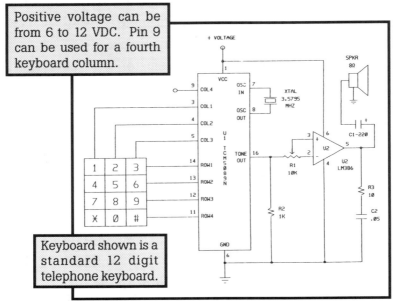

Figure 22-1. TCM5089N based touch-tone generator.

Project #22: Touch-Tone Generator

generate two different audio tones whose combination represents either a numeral or a symbol. Both are in the audio range, with one tone high and the other low (see chapter two).

The reason for the dual tones is to prevent the possibility of telephone line noise causing false signals. It wouldn't be extraordinary for a stray tone of the proper frequency to appear on the line, but the prospect of both tones appearing at exactly the same time is considerably less likely. Thus, the dual tone arrangement provides excellent protection against such noise.

So, how do you produce these all important tones? The answer is the DTMF tone generator. Actually this can be done with the LM555 timer/ oscillators and an elaborate switching network, but the dual tone chips are far better (easier). There are several different ICs that will accommodate us, but I have chosen the TCM5089N integrated tone dialer because of availability and cost.

Unfortunately, chips of this nature can come and go rather quickly, but the 5089 has been a standard for several years, and probably will remain so. Remember, I promised at the outset to avoid circuits using obsolete, or "here today, gone tomorrow" ICs, and this is a conscientious effort to keep that promise. Of course, there are no absolutes, and by the time you read this the TCM5089 may be just a fond memory. At this point in time, however, it is alive and well.

You'll note from **Figure 22-1** that the chip has eight lines dedicated to the keyboard. Four of these address the columns, while the remaining four handle the rows. For the standard telephone keypad (12 keys) the column 4 pin is not used, although it would be needed if you plan on a sixteen key pad with A, B, C and D switches.

For most telephone oriented systems, 12 keys is all that is required, thus U1's pin 9 is left open, or floating. Pins 7 and 8 accommodate a 3.579545 megahertz color burst quartz crystal that provides circuit timing. The clock is, of course, an on-board feature of the TCM5089.

The output, pin 16, delivers the tone pair, and in our configuration is connected to the input of an LM386 audio amplifier. The purpose of the amp is to boost the output to speaker level so the tones can easily be heard by you, or the telephone sound pickup. This stage is the standard layout for the LM386.

Power for the circuit can come from a 6 to 12 volt battery, a regulated AC supply, or the telephone line. If the AC option is chosen, be sure it is well filtered to prevent 60 hertz hum interference. Naturally, a battery will not present such a problem.

CIRCUIT DESCRIPTION

The purpose(s) you have planned for this device will play a significant role in how you construct the project. As always, the circuit can be breadboarded for testing, and if you want it to be portable, such as a touch-tone dialer for non-tone mode phones, then a PCB is recommended.

If, however, it will be work bench equipment, then either construction style will do quite nicely. For this second option, an AC power supply will probably be more convenient.

A third possibility would be to modify a pulse mode telephone to touch-tone, in which case you will need to power the device from the phone line itself. This is not difficult. However it does require some precautions.

A rectifier bridge is necessary to protect the circuit from polarity reversals on the telephone line, a metal oxide varistor (MOV) is recommended for surge protection, and a zener diode will regulate the operating voltage. **Figure 22-2** shows an accept-

Project #22: Touch-Tone Generator

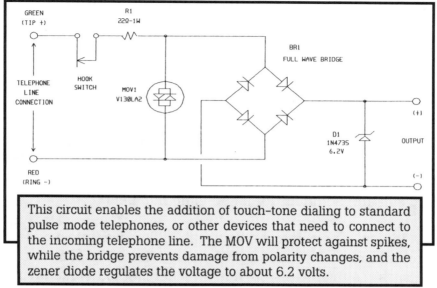

Figure 22-2. Telephone line interface - DTMF generator.

able power circuit. Installation of this circuit will not only provide power, but the needed safeguards as well. Only the red wire (RING -), and the green wire (TIP +) are used for the interface to the telephone system, so modifying an older phone is not terribly complicated.

The size of the speaker depends on your intended application. For a pocket dialer, a small one inch job will do fine. With a bench design, you will probably want to use a larger speaker. Additionally, a jack may be desirable for direct connection to devices such as the DTMF decoder in chapter two.

Potentiometer R1 is the volume control, and again, the style of equipment in mind will influence your approach to this component. For the portable version, a PCB type potentiometer will serve nicely, as you want the volume loud enough to activate the telephone company's central office equipment, but not so loud as to blast you. However, once set, it will remain there, barring extenuating circumstances.

Regarding other configurations, you may well want this to be panel mounted for adjustable volume. This is primarily a matter of personal preference and/or perceived utilization.

CONCLUSION

The DTMF concept was really a stroke of genius on the part of AT&T. Not only did it modernize telephone dialing, but also made it far more reliable. The old fashion rotary phones were, and still are, notorious for mis-dialing numbers.

In addition, tone pair technology has spilled over into a variety of other areas. One arena profoundly effected is control circuits, as the dual tones provide an ideal method of managing all sorts of stuff. Ask any amateur radio operator (HAM) about DTMF control of repeaters.

The telephone system itself uses the tones not only to dial numbers, but to direct the incoming calls to specific departments or persons. I know this is sometimes annoying, but don't blame the technology. It works!

The system also allows you to call your home and start/stop various appliances, check answering machines, and even monitor any unusual activity via listening to sounds within the dwelling. This takes special equipment, of course, but all of it is based on the principle of DTMF.

So, with this generator you can get a first hand look at how DTMF functions, and perform some interesting experiments and demonstrations. Don't forget chapter two, and the DTMF detector. It will tell you exactly which tone pair is being generated. That is, what number or symbol the pair represents. All in all, I think you are going to find numerous uses for this device. If nothing else, you can always play your favorite tune on it.

Project #23: Frequency to Voltage Converter

What You Will Need:

- ✔ U1   LM2917, Frequency to Voltage Converter
- ✔ D1   1N4001 Rectifier Diode
- **All resistors 5% carbon film.**
- ✔ R1, 4   10,000 Ohm, 1/4 Watt Resistors
- ✔ R2   470 Ohm, 1/4 Watt Resistor
- ✔ R3   33,000 Ohm, 1/4 Watt Resistor
- ✔ C1, 2   0.02 Microfarad Mylar Capacitors
- ✔ Copt   0.1 Microfarad Mylar Capacitor to (+) Input (see text)

Integrated Circuit Projects

Some years back, National Semiconductor introduced the LM2907/LM2917 series of frequency to voltage converters. Originally designed for automotive use, they are, to this day, an excellent means of performing that conversion. Additionally, both chips can be utilized in a number of other areas.

National's application notes suggest over/under speed sensing, tachometers, speedometers, breaker point dwell meters, speed governors, cruise controls, automotive door lock control, clutch control, horn control, and touch or sound switches.

The two ICs are practically the same, but the 2917 has some minor improvements. Thus, this is the chip we will work with. It comes in either an 8 pin or 14 pin package, and I have selected the 14 pin, variety. Oh, one other point, it's cheaper than the 2907.

While the data sheet doesn't give frequency parameters, my experience is that the LM2917 will reach 15 to 20 kilohertz without any problem. Thus, it can also be used in audio frequency meters. However, with some thought, and keeping that restriction in mind, I'm confident you can find even more applications for this IC.

CIRCUIT THEORY

Essentially, the LM2917 consists of three parts; a high gain op-amp, a comparator, and a tachometer. The op-amp and comparator are combined in one segment, and can be used to trigger relays, lamps, or other loads.

The tachometer section employs a charge pump arrangement to do its work, and this provides frequency doubling for low ripple, an output swing to ground for an input of zero, and input protection. This part of the IC allows for the speedometer and tachometer circuits, while the op-amp/comparator handles over/under speed sensing, speed governors and the like.

Project #23: Frequency to Voltage Converter

The various control functions (horn, clutch, etc.) would also be a product of the op-amp stage, but we will be looking at a tachometer design, as this will meet our needs. By detecting changes in frequency, this circuit can be employed in any configuration that requires such detection. Keep the frequency limitation in mind.

Figure 23-1 illustrates the basic circuit for this type of application. Note the small number of discrete, or external, components required, as the 2917 does most of the work. Even better, those discrete parts are commonly available off-the-shelf items.

Specifically, the LM2917N is a 14 pin standard DIP with a maximum rating of 28 volts. That puts it in range of just about any design. It draws around 25 milliamps, and this allows for portable operation, if needed. One precaution is not to permit an input voltage to exceed the operating voltage, but that is common practice with all ICs.

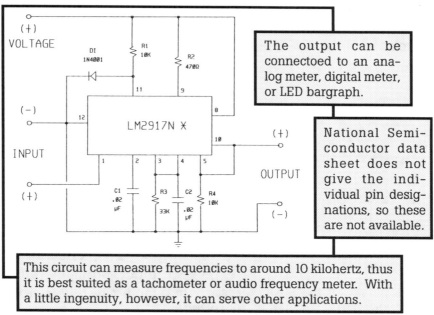

Figure 23-1. Frequency to voltage converter.

One exception to that rule is the tachometer, which is protected against an accidental swing above the operating voltage. Bear in mind this protection only applies to a momentary overage, not a continuous excess.

Another handy characteristic of the 2917 is that it will interface directly with magnetic pickups commonly used for reciprocal engine tachometers. This resolves one often problematic aspect of designing this type of device.

Additionally, the 50 milliamp output provides ample current for meters, both analog and digital, relays, solenoids, or LED. Here again, another obstacle is overcome by the IC itself.

CIRCUIT DESCRIPTION

Viewing **Figure 23-1**, one cannot help but appreciate the simplicity of this circuit. Simplicity to the point that the entire unit will easily fit on a piece of construction material a couple of inches square. This can be perf-board, or a PCB, whichever you prefer.

Layout is not critical, but mylar style capacitors are recommended for their exceptional dielectric characteristics. Other than that, there are no special considerations to observe, just good technique.

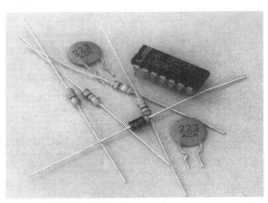

Photo 23-1. All the parts necessary to build the 2917 frequency to voltage converter.

Mentioned in the parts list is capacitor *Copt*, and this refers to a 0.1 microfarad mylar that may be needed for the input. Normally, interfacing is not troublesome, but if problems

Project #23: Frequency to Voltage Converter

are encountered, placing this capacitor in series with the positive input (pin 1) will usually resolve any difficulties. However, don't use it unless you have to, as most of the time it is just excess baggage.

Once you have the basic circuit constructed, it is time to put it to work. Let's stay with the automobile tachometer concept, for this will illustrate the chip's functionality.

The magnetic pickup can be purchased at most automotive supply stores for a few dollars, and the instructions will show you how to install it on your model vehicle. With it in place, the leads are now connected to the input of the frequency to voltage converter.

Any type of wire will work, but some form of shielded cable, maybe audio coax, is recommended. This will prevent false readings resulting from the abundance of electrical noise under a hood. At this time you also want to hook up the power connections. Since the tachometer will only be needed when the engine is running, it is best to hook the power cable to a source that is hot just when the vehicle is operating.

Now comes the output, and it needs to be connected to a visual indicator. Depending on your preference, this can take the form of the traditional analog meter, or for a more modernistic look, a digital readout or LED bargraph will fill the bill.

Each has advantages and disadvantages, but regardless of which annunciator you choose, the magnetic pickup will detect the number of engine revolutions (frequency), and present this to the 2917. The IC will, in turn, convert that to a voltage and drive the gauge. A digital meter is probably more precise, but you may lean towards the colorful LEDs, or the nostalgic appearance of a needle moving over an analog scale.

CONCLUSION

A home built automobile tachometer is a very practical application of the LM2917 frequency-to-voltage converter. Not only is this an interesting project, but it will save you big bucks considering the price of commercial gauges.

An excellent digital display is the RCA voltmeter we built in chapter fourteen. If you hardwire the first, or least significant digit, as a zero (a, b, c, d, e, and f all tied to ground), then the meter can act as the three most significant digits.

A bargraph arrangement can be achieved by modifying the circuit in chapter one. You will, however, want to substitute an LM3914 for the LM3916. In either case, LEDs tend to wash out in bright sunlight, so try to mount the display in a shaded position on the dash.

Well, now you're ready to watch that engine performance and save that gas. The last point is mighty important these days, unless you have a generous uncle who owns a significant portion of Saudi Arabia. Anyway, don't forget the versatility of this chip. Try your hand at developing other devices based on the 2917, as I think you will discover it wears many a pragmatic hat.

Project #24: Infrared Transmitters and Receivers

What You Will Need:

Infrared Transmitter:

- ✔ LED1 — Jumbo T-1 3/4 Infrared Light Emitting Diode
- ✔ R1 — 1,000 Ohm, 1/4 Watt Resistor (5% carbon film)
- ✔ B1 — 9 Volt Alkaline Battery
- ✔ S1 — Momentary Contact, Normally Open Push Button Switch

IR Receiver:

- ✔ Q1 — Infrared Sensitive Phototransistor (any size)
- ✔ Q2 — 2N3904 NPN Silicon Transistor
- ✔ D1 — 1N914 Signal Diode
- ✔ R1 — 100,000 Ohm Potentiometer
- ✔ RL1 — PCB Style Relay (voltage according to VCC)

Integrated Circuit Projects

IC Based IR Audio Receiver:

- ✔ U1 — TL071CP, Low Noise JFET Operational Amplifier
- ✔ U2 — LM386N-3, Low Voltage Audio Amplifier 500mw/9V
- ✔ Q1 — Infrared Sensitive Phototransistor (any type)
- ✔ R1, 2 — 100,000 Ohm, 1/4 Watt Resistors
- ✔ R3 — 10,000 Ohm Potentiometer
- ✔ R4 — 10 Ohm, 1/4 Watt Resistor
- ✔ C1, 2 — 0.1 Microfarad Mylar Capacitors
- ✔ C3 — 220 Microfarad Electrolytic Capacitor
- ✔ C4 — 0.05 Microfarad Mylar Capacitor
- ✔ SPKR — 8 Ohm Speaker

Modular Receiver:

- ✔ Q1, 2 — 2N3904 NPN Silicon Transistors
- ✔ MOD1 — Infrared Receiving Module (RS, Jameco, Digi-Key)
- ✔ R1, 2 — 1,000 Ohm, 1/4 Watt Resistors
- ✔ R3, 4 — 10,000 Ohm, 1/4 Watt Resistors
- ✔ R5 — 4,700 Ohm, 1/4 Watt Resistor
- ✔ C1, 2 — 0.1 Microfarad Mylar Capacitors
- ✔ T1 — 8 Ohm to 1.2 Ohm Audio Transformer (Jameco)
- ✔ SPKR1 — 8 Ohm Speaker

Infrared Voice Transmitter:

- ✔ U1 — LM386N-3, Low Voltage Audio Amplifier 500mw/9V
- ✔ LED1 — Infrared Light Emitting Diode (any type)
- ✔ R1 — 1,000 Ohm, 1/4 Watt Resistor
- ✔ R2 — 10,000 Ohm Potentiometer
- ✔ R3 — 10 Ohm, 1/4 Watt Resistor
- ✔ R4 — 470 Ohm, 1/4 Watt Resistor

Project #24: Infrared Transmitters and Receivers

- ✔ C1   0.2 Microfarad Mylar Capacitor
- ✔ C2   0.05 Microfarad Mylar Capacitor
- ✔ C3   220 Microfarad Electrolytic Capacitor
- ✔ MIC   Electret Microphone

Increased Sensitivity Voice Transmitter:

- ✔ U1   LM386N-3, Low Voltage Audio Amplifier 500 mw/9V
- ✔ Q1   2N3904 NPN Silicon Transistor
- ✔ LED1   Infrared Light Emitting Diode
- ✔ R1   1,000 Ohm, 1/4 Watt Resistor
- ✔ R2   100,000 Ohm, 1/4 Watt Resistor
- ✔ R3   10,000 Ohm, 1/4 Watt Resistor
- ✔ R4   10,000 Ohm Potentiometer
- ✔ R5   10 Ohm, 1/4 Watt Resistor
- ✔ R6   470 Ohm, 1/4 Watt Resistor
- ✔ C1   2.2 Microfarad Electrolytic Capacitor
- ✔ C2   0.05 Microfarad Mylar Capacitor
- ✔ C3   220 Microfarad Electrolytic Capacitor
- ✔ MIC   Electret Microphone

The invention of the infrared remote control really brought this form of light into its own. Before that, most people thought of infrared as an orange colored flood lamp that heated stuff, or gave you a tan. Indeed, it is all of that, but so much more as we will see in this chapter.

The infrared band of light (IR) is comparatively wide, and found off the end of the red light spectrum. Today, IR is easily produced by lasers and LEDs, as well as the conventional incandescent bulbs. However, its value lies more in the field of communications than keeping the burgers warm at your run of the mill fast food joint.

The ability to modulate infrared light makes it nearly as versatile as RF energy, and the proof is all around us. IR remote controls for televisions, video recorders, stereo equipment, as well as lamps, fans, computer mice, video games and the like, all stand as a monument to infrared's flexibility.

But, that is just the tip of the iceberg. Telephone, radio, and even television are areas that not only use infrared, but continue to conduct large research programs in an effort to improve its application. From what I hear, here and there, the research is not in vain.

So, with that preface under your belt, let's look at some infrared circuits. Most of these would be considered basic in nature, but they will provide a foundation for your investigation of this field. We will inspect both transmission and reception circuits for both single action devices (switches) and voice (sound) communications.

CIRCUIT THEORY

In this section, I want to provide you with general theory, much of which we have already touched on in previous chapters. Hopefully, in this fashion you will get a feel for how infrared light can be employed in a control capacity, as well as data transfer.

Modulation is modulation. What this means is that by altering the intensity, or frequency, of radio waves, light waves, and so forth, you can transmit information. In the end, it, doesn't matter which medium (carrier) you use, as the result is the same. The information is carried from one point to another.

Thus, your choice of medium depends more on circumstances, necessity and convenience than anything else. Light, of course, will not penetrate a solid wall, so if your application requires that, it is not practical. However, it can be easily focused for

Project #24: Infrared Transmitters and Receivers

precise aiming. It isn't affected as badly by extraneous interference as RF, and circuit-wise, can be far less complex than radio. Note that I say can be, as advanced IR systems often get quite complicated.

So, for short range, line of sight operations, light is a natural. It provides a stable carrier, excellent reliability, and in the case of infrared, is just as transparent as RF. A light beam can transport that modulation just as well as other forms of energy.

Once the infrared is transmitted, be it modulated or not, the receivers will detect its presence and trip a relay to turn something on or off. In the case of voice communications, the beam is un-modulated to recreate the original message. With multi-function remote controls, modulation takes the form of a serial signal, or series of short pulses. Again, this is decoded by the receiver, and each sequence of short pulses indicates a needed operation (turn up the volume, change the channel, etc.).

As will be seen in the next section, the detectors can be one or more IR sensitive phototransistors, or an infrared module that contains a whole parcel of electronic devices. These greatly improve the data transmission, and the modules have become a staple in the industry for their compact size, but high performance characteristics.

CIRCUIT DESCRIPTIONS

For this chapter, I have included six separate circuits that should give you an overview of infrared technology. As stated earlier, these are basic configurations, and many IR systems are substantially more complex, but in most cases the principle is the same.

Let's examine each of the circuits, one at a time. Starting with **Figure 24-1A**, this is an infrared transmitter in its most primi-

Integrated Circuit Projects

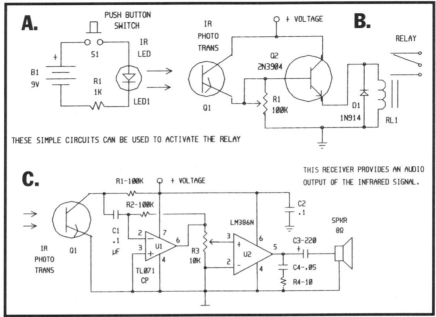

Figure 24-1. *A: IR transmitter. B: IR receiver. C: IC based IR audio receiver.*

tive form. When switch S1 is pushed, resistor R1 drops battery B1's current to a safe level, and lights the IR LED. This produces a continuous output of infrared light.

A receiver, such as the one in **Figure 24-1B**, employs an infrared phototransistor to detect the light, and uses driver transistor Q2 to trigger the relay. The relay can then turn any number of devices on and off, depending on the state of the transmitter. Simple, but effective.

The problem with this type of system is that virtually any source of infrared light will activate the receiver. Thus, changing the settings on the TV or stereo will also energize this gem. This is because the infrared beam is not encoded (modulated is some fashion), and the receiver does not employ decoding (interpreting a correctly coded signal).

Remember, the remote controls use serial pulses to define what action they want taken, and a similar arrangement could be

Project #24: Infrared Transmitters and Receivers

incorporated in this design, making it immune to extraneous IR light. Holtek has developed an extensive line of encoder/decoder chips, and these would do nicely for modifying either the transmitter or receiver.

Additionally, a silicon control rectifier (SCR) could be used to lock and unlock the relay each time the switch S1 is pressed. That would eliminate the need of holding the switch down to keep the receiver activated. Hence, there are many ways to modify the original circuits and provide extra versatility.

Figure 24-1C illustrates the first of two infrared sound receivers. Both are designed to decode the modulated beam sent by a transmitter, and as seen, this unit utilizes a phototransistor as the detector. Once the signal is picked up, it is sent to an op-amp for buffering and pre-amplification, then on to an LM386 for power amplification and speaker output.

While only two ICs strong, this receiver will provide amazingly good results. The detector's (Q1) sensitivity can be increased with a parabolic reflector and/or optics, and this will improve the range (see **Figure 24-3**). R3 acts as the volume control, and a jack could be included for headphones.

Figure 24-2A shows a second receiver, and this one employs an infrared module as the detector. Instead of ICs, this time we use transistors to first invert the module's output (Q1), then amplify the signal (Q2). Since transistors produce a high impedance output, audio transformer T1 is required to step that impedance down to 8 ohms. This will match the speaker and avoid distortion.

These modules may not look like much from the outside, but inside they are something to behold. Let's take a closer gander at what they do. **Figure 24-4** is a block diagram of the circuitry inside the module, but why so complicated?

Integrated Circuit Projects

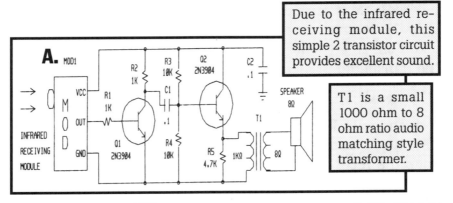

A.

Due to the infrared receiving module, this simple 2 transistor circuit provides excellent sound.

T1 is a small 1000 ohm to 8 ohm ratio audio matching style transformer.

Using an LM386 audio amplifier chip, the electret microphone will detect sound which will be converted into a change in intensity of the infrared LED. This is known as encoding, and the receiver in turn decodes the changing light to reproduce the original sound.

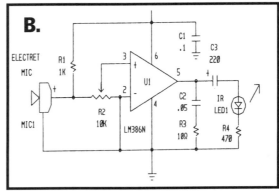

Figure 24-2. A: Modular receiver. B: Infrared voice transmitter.

Well, initially, most IR devices send their data on a 40 kilohertz carrier, and we want to maintain that. A photodiode, instead of a phototransistor, is used to detect the IR signal, and that is sent to a high gain op-amp. In order to convert the amp's output to a semi-square wave, a limiter is employed to chop off the top of the sine waves.

Next, a bandpass filter is used to narrow the frequency to about 40 kilohertz +/- 4 kilohertz. This removes noise, and prevents ambient infrared of a different frequency from affecting the reception. This somewhat conditioned input signal is now rectified and run through the integrator, which is another filter designed to react slowly to the signal.

Project #24: Infrared Transmitters and Receivers

You want this to be too slow to keep up with the 40 kilohertz carrier, but not so slow that it misses the pulses of the original signal. The purpose here is to isolate the code from the carrier, which is important to the next stage.

That stage is called the comparator, and is actually a Schmitt Trigger. If you are familiar with the Schmitt Trigger, then you know that a certain signal amplitude has to be reached for the trigger to react. That is precisely what we want, as only the code pulses, and not the 40 kilohertz carrier, should have a high enough amplitude to activate the comparator.

This last stage does invert the signal, however, so a simple transistor inverter (Q1 in **Figure 24-2A**) is used to re-invert it to what is needed. Without the re-inversion, the circuit would decode the opposite, or mirror image of the original transmission. That would never do, especially for audio.

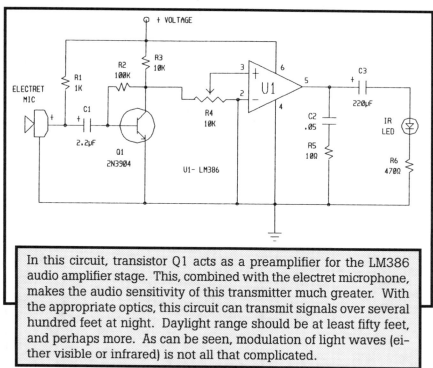

In this circuit, transistor Q1 acts as a preamplifier for the LM386 audio amplifier stage. This, combined with the electret microphone, makes the audio sensitivity of this transmitter much greater. With the appropriate optics, this circuit can transmit signals over several hundred feet at night. Daylight range should be at least fifty feet, and perhaps more. As can be seen, modulation of light waves (either visible or infrared) is not all that complicated.

Figure 24-3. Increased sensitivity voice transmitter.

So, you see I wasn't kidding. There is a lot under the hood of an infrared module. About 300 horsepower I would say. Of more importance, though, is the performance you get out of these gems. While I have shown it in an audio receiver capacity, the module will do an admirable job in all types of infrared circuits.

Ok, let's turn our attention to the voice transmitter circuits. **Figure 24-2B** shows a very simple design that employs the LM386 audio amplifier chip. In this configuration, the electret microphone picks up your speech which is amplified by U1. The output of the IC is then used to vary the intensity of the infrared LED, with the loudest sound giving the brightest illumination.

Bear in mind that this is infrared, so you won't be able to see this variation. However, if you want to temporarily connect a standard LED, you will be able to watch it flicker in response to the audio input. This, by the way, affords an excellent precheck of proper circuit operation.

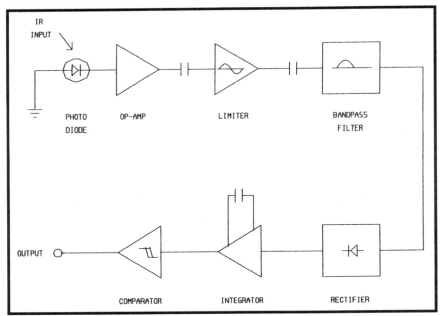

Figure 24-4. IR module block diagram.

Project #24: Infrared Transmitters and Receivers

The design provides reasonably good sensitivity, but that can be improved through the addition of a pre-amplifier stage. Figure 24-3 illustrates just such a circuit, with transistor Q1 acting as a single stage pre-amp. This boosts the microphone pickup before introducing it to the 386 power amp, thus making the unit far more sensitive to low sound levels. You may well recognize this transistorized pre-amp from chapter eleven.

There you have it. Transmitters, receivers, and switching circuits, all of which can be put to conscientious use. I think you will be surprised at the performance of the voice units. With a little bit of detector shielding (a cardboard tube around the phototransistor to keep out ambient light), you should be able to get ranges of fifty feet or better in daylight conditions, and several hundred feet at night.

Again, with the addition of parabolic reflectors, optical collimators, or both, those ranges can increase vastly. All in all, it isn't hard to set up a very respectable full-duplex voice link with a transmitter and receiver at each end. Considering the simplicity of the electronics, it wouldn't be expensive either.

CONCLUSION

In my years as an experimenter and hobbyist, I have seen the birth of ICs, microprocessors and LEDs, and each one brought a new sense of awe for me. Virtually all of this has come to pass since about 1960, and there seems to be no end to further developments.

As I think back over that time, there are a few discoveries that really grabbed my attention. The laser was one, and the LED was another. Each displayed, at least in my mind, science at its best, and both illustrated the ingenuity mankind could muster when put to the test.

Integrated Circuit Projects

Thus, this chapter represents an area of special interest for me; the application of the LED in a capacity other than a pretty red glow on the front panel. A device that can actually send data of all types, over considerable distances, on a simple beam of invisible light.

Over the years, a few have mocked me, but I can't help it. These concepts are, to this day, exciting. Bearing such imagination, and the talent to transform it into tangible reality, man is destined to yet reach out and touch the stars.

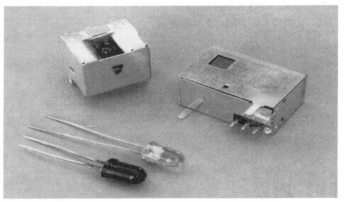

Photo 24-1. Top: Two typical infrared modules. Bottom: Infrared matched pair of LED and detector.

What You Will Need:

✔	U1	ZN414Z, AM Radio Receiver to 3 MHz
✔	U2	LM386N-3, Audio Amplifier 500mw-9V
✔	R1	100,000 Ohm, 1/4 Watt Resistor
✔	R2	10,000 Ohm Potentiometer
✔	R3	10 Ohm, 1/4 Watt Resistor
✔	C1	150 to 365 Picofarad Variable Capacitor
✔	C2	0.01 Microfarad Mylar Capacitor
✔	C3	0.1 Microfarad Mylar Capacitor
✔	C4	220 Microfarad Electrolytic Capacitor
✔	C5	0.05 Microfarad Mylar Capacitor
✔	L1	Ferrite Tuning Coil (50 to 60 turns on ferrite rod)
✔	SPKR	8 Ohm SPeaker
✔	A1	DPST Toggle, Slide, Other Switch
✔	B1	1.5 Volt Battery
✔	B2	9 Volt Battery

Integrated Circuit Projects

AM radio just isn't as popular as it was twenty years ago. There is still a lot to hear on that band, from small budget music stations to the sometimes controversial, politically oriented talk radio programs.

But that is not the purpose of this chapter, at least I hope not. No, what I want to talk about here is a chip that allows you to build an old fashion AM broadcast band radio very simply. I mean, with the possible exception of a crystal set, it doesn't get any simpler than this.

At the center of this project is the ZN414Z AM radio receiver to 3 Megahertz IC. This is an ingenious device that comes in a TO-92 three pin package, the same size as many transistors. With the addition of a few external components, you have an AM band receiver front end. Add an amplifier, and you have a complete radio.

Needless to say, considering the size of the ZN414, the whole circuit can be kept quite small. The major obstacles to compactness are the ferrite tuning coil and variable capacitor, as these have to be pretty much standard sized components.

However, a pocket AM radio can easily be constructed from the ZN414, and it takes a lot of the problematic aspects out of that construction. Once again, modern electronics comes to the aid of the aspiring hobbyist, making his or her task more enjoyable.

CIRCUIT THEORY

For starters, let's take a close look at the ZN414 itself. **Figure 25-1** is a block diagram of this chip, and as can be seen, a bunch of stuff has been packed into this tiny IC. In all, ten transistors and dozens of discrete components occupy the interior, resulting in a complete tuned radio frequency (TRF) configuration.

Project #25: ZN414Z Based AM Radio

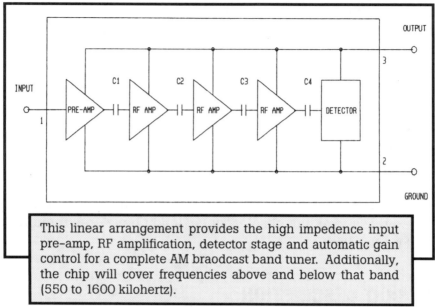

Figure 25-1. ZN414Z internal block diagram.

As **Figure 25-1** reveals, the L/C derived input is fed to a pre-amp, then sent to three RF amplifier stages. Finally, the signal is introduced to a detector section that produces the output. Believe it or not, all of this fits inside that tiny TO-92 transistor case. The completed circuit is remarkably stable in operation. Additionally, an automatic gain control (AGC) is included for even better performance. This, too, is neatly stashed in the little plastic package.

In its simplest form, only capacitors C2 and C3, resistor R1, and a low impedance crystal earphone are required for operation. However, the volume isn't very loud, so we add an audio amplifier stage (U2) to eliminate this problem.

You guessed it, the old stand-by LM386 is, again, pressed into service, and an admirable job it does. Utilizing its usual simple configuration, the 386 provides around 500 milliwatts of output giving the ZN414 its voice.

One drawback to the chip is the 1.5 volt requirement. While at first glance this might seem advantageous, it does present a power problem for the rest of the circuit. The LM386 needs at least 6 volts to operate, but 6 volts would fry the 414. Thus, two batteries, or power supply voltages, are essential.

Figure 25-2 illustrates a layout that accommodates this necessity, so it really isn't all that inconvenient. I am told that at one time, a zener diode with the designation KS047A was available. This would regulate the higher 9 volts down to 1.5 volts, but I honestly could not tell you where to find one today. I'm afraid it is a victim of obsolescence. If you have one, though, it will do the trick.

CIRCUIT DESCRIPTION

This circuit is the essence of efficient simplicity, and that makes it a natural candidate for educational purposes, as well as just plain fun. The hardest part about this project will be finding the tuning coil and capacitor, but never fear, I have some ideas for you.

The capacitor can be obtained from several sources, see the parts list, and the coil can either be hand wound or robbed from an old AM radio. For that matter, the obsolete radio can supply the capacitor as well.

PCB construction provides a tidy finished product, but point-to-point wiring will work just as well. That is more a matter of personal preference than anything else.

Only variable capacitor C1 and volume control potentiometer R2 require access, so that further simplifies construction. Actually I lied. Power switch S1 also has to be kept handy to turn the thing on. Other than that, everything else can be out of sight.

Project #25: ZN414Z Based AM Radio

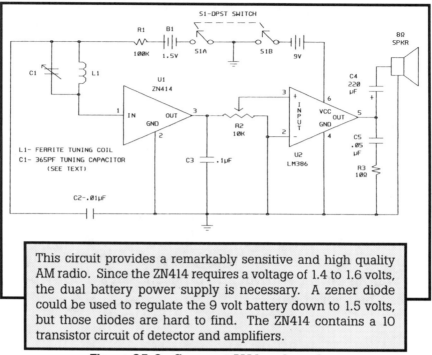

This circuit provides a remarkably sensitive and high quality AM radio. Since the ZN414 requires a voltage of 1.4 to 1.6 volts, the dual battery power supply is necessary. A zener diode could be used to regulate the 9 volt battery down to 1.5 volts, but those diodes are hard to find. The ZN414 contains a 10 transistor circuit of detector and amplifiers.

Figure 25-2. Compact AM band receiver.

I mentioned that tuning coil L1 can be hand wound, so let's talk about that. This is not hard. Find a 1/4 to 3/8 inch ferrite rod a couple of inches long, and wind 50 to 60 turns of small enameled wire (#28, #30 or #32) on it. Coat the finished product with clear nail polish, let it dry, and you have your coil. Don't forget to remove the enamel from the ends of the lead wires before soldering them.

And that is about all there is to it. The discrete components are standard items, and there is nothing particularly sensitive, crucial, critical, difficult or frightening about the overall circuit layout. Use your normal expertise, and have at it.

CONCLUSION

Using the ZN414Z is a real treat. I have built several receivers with this chip, and I always marvel at the outcome. Once I helped a friend design a monitor for a low power (licence free) beacon that operated at 1730 kilohertz (this is something that the amateur radio people do), and I remember her surprise at my suggesting this IC.

That surprise was only surpassed by her surprise at the superb performance of the monitor. It seemed the beacon was over four hundred miles away, and hard to hear on a shortwave receiver, but it came booming through on the ZN414 design.

Although I can't personally verify this, I have been told it will drop down into the very low frequency (VLF) range as well, and there is also some amateur radio activity down there. This is an area where you can legally operate transmitters up to one watt in power without an FCC licence. It's a noisy band, but can be a lot of fun.

Conclusion

Much of the electronics industry has its roots in the garage labs of experimenters. Throughout its legacy, names like Nicola Tesla, Alexander Bell, Heinrich Hertz, Frederick Collins, H.J. Round, Lee De Forest, Samuel Morse, Philo T. Farnsworth, and numerous others have been irrevocably tied to the development of electronics. Some names you may recognize, some not, but many of these people were garage experimenters, and not engineers.

This is not to downsize the importance of engineers. Indeed not, as they too have contributed admirably to the field, especially in the area of semiconductors. The point being made here is that hobbyists and experimenters, like yourself, are responsible for immense advancements. Philo Farnsworth was only sixteen when he laid the foundation for modern day television, only to have it stolen by a major American corporation.

Be that as it may, experimenters remain undaunted in their efforts to expand the horizons of electronics. Let it be a lesson, though. Guard your ideas, but don't stop exploring them. For without you, the industry would hardly be what it is today.

Unlike the engineers, for whom this is *work*, for us this is *fun*, and that can be a huge advantage. I remember my hobby of photography, that one day turned into labor in the form of a photojournalism job. After that, photography was never quite the same. It always had the taint of work. I still enjoyed it, and do to this day, but somehow it has lost the blithe aspect it once held for me.

Fortunately, for me electronics has never been anything but fun. Although it eventually became work, I avoided that pitfall by staying with the hobby/experimental side of the subject. Here, I found a world of fascinating voyages into all areas of this scientific sphere, and a world not as complex as some would like you to believe.

Additionally, that world changed with such briskness that you were never without a new horizon to explore. The advent of transistors and integrated circuits further expanded that horizon, and so I was hooked. The knowledge was enticing, and the results took the form of practical working devices. What more could you ask for?

Then came the ultimate enjoyment. One day I found myself with the faculties to design my own circuits, and the sky was the limit. Now, when I needed a special piece of equipment, I summoned the knowledge and experience of my years as an experimenter, and quickly drew up plans for my project. I have to tell you, that first working prototype was a hoot.

So, if a dummy like me can do it, it will be easy for you. All that's required is the dedicated interest and inquisitive nature that lives in all of us. The rewards are both tangible and intangible, and very abundant. But, I don't need to tell you that, as your reading this book would indicate you have already made that discovery.

I'm glad you are, though, for hopefully it will assist you in your pursuit of all that electronics can provide. If you are just starting out, so much the better. You have ahead of you the very best days this hobby can offer. The knowledge and ability will come at a surprising rate, bringing with it satisfaction, pride and enjoyment.

Conclusion

For the old hands, I hope this text provides something new, and perhaps a handy reference source of circuits. Two books are better than one, you know. Well, at least that has been my experience.

In closing, let me again thank you for your interest in electronics. It does make a big difference in the future of the field. Believe it or not, there was a time when Farnsworth's type of television was in serious jeopardy. A substantial number of people thought TV should be based on mechanical spinning disks with little holes in them, and they almost got their way.

Had it not been for the perseverance of advocates of the electronic method, you would probably be watching a box that produced images resembling a bad home movie. That's if television had survived at all. So, keep the faith. Your interest and support is what perpetuates the industry. Besides, it keeps me in work.

Most of all, though, have fun with it. The technology is on the fast track, and producing new and invective components, devices and approaches every day, every hour, every minute. So much so that it is impossible to keep up with everything, but I try. I try because each new concept affords another avenue of experimentation and exploration.

Integrated Circuit Projects

Appendix A

This is a list of some of the many electronic parts suppliers serving hobbyists and experimenters. I have included these companies in alphabetical order because, over the years, they have proved themselves to be reliable, customer service oriented, and reasonably priced. However, you may well have favorites of your own.

ALL ELECTRONICS CORPORATION

P.O. Box 567
Van Nuys, CA 91408
Order Line: (800) 826-5432
Customer Service: (818) 904-0524
FAX: (818) 781-2653

This company carries a good line of general electronic parts, tools, and hardware. They also handle some interesting hobby and surplus items, as well as a few kits.

ALLTRONICS

2300 Zanker Road
San Jose, CA 95131
Customer Service: (408) 943-9773
FAX: (408) 943-9776
BBS: (408) 943-0622

Integrated Circuit Projects

Extensive selection of surplus electronics from parts to computer peripherals and other equipment; some new, some used. Also carries one of the best lines of out-of-production ICs around. A little high on shipping/handling charges.

AMERICAN SCIENCE AND SURPLUS

3605 Howard Street
Skokie, IL 60076
Customer Service: (847) 982-0870
FAX: (800) 934-0722

Excellent inventory of new and used surplus items. Included is hardware, lab apparatus, office supplies, gadgets, mechanically oriented parts and devices, optics, lasers, tools, and of course, electronics. You never know what treasures you will find in their catalogs, which come about every two months.

B. G. MICRO

P.O. Box 280298
Dallas, TX 75228
Order Line: (800) 276-2206
Tech. Support: (214) 271-9834
FAX: (214) 271-2462

Carries computer items, parts, ICs, novelty electronic equipment, surplus, new and used, and a good line of kits. Interesting catalog, with good prices and fast service.

CIRCUIT SPECIALISTS, INC.

P.O. Box 3047
Scottsdale, AZ 85271-3047
Cust. Service/Order Line: (800) 528-1417
FAX: (602) 464-5824

An extensive line of computer products, boards, and systems. In addition, they carry kits, test equipment, and a substantial inventory of components and prototype materials, as well as educational, and fiber optic supplies. Prices competitive.

Appendix A

DC ELECTRONICS

P.O. Box 3203
Scottsdale, AZ 85257-3293
Order Line: (800) 467-7736
Order Line: (800) 423-0070
Local: (602) 945-7736
FAX: (602) 994-1707

DC carries a respectable line of discrete components, but excels in the semiconductor area. Many hard-to-find ICs, and transistors are listed in their catalog. They also have an excellent line of PCB and prototype supplies. "TINNIT" and TEC-200 Image Film are available from this company.

DEBCO ELECTRONICS, INC.

4025 Edwards Road
Cincinnati, OH 45209
Order Line: (800) 423-4499
Information: (513)531-4499
FAX: (513)531-4455
E-mail: debco@iglou.com
Internet: http://eli.warait.org/debco-el/index.html

Good selection of kits, components, ICs, tools, and miscellaneous electronic items. No minimum order, and their catalog, they call the "Electronic Experimenter's Journal", includes some interesting and helpful basic electronic circuits and information.

DIGI-KEY CORPORATION

701 Brooks Avenue
Thief River Falls, MN 56701-0677
Cust. Service/Order Line: (800) 344-4539
FAX: (218) 681-3380

Handles a huge inventory of current components, hardware, tools, and equipment. Good source of "PIC" microcontroller products. This company stays on top of technological advances, but is not very helpful when older or out-of-production supplies are needed.

ELECTRONIC GOLDMINE

P.O. Box 5408
Scottsdale, AZ 85261
Customer Service: (602) 451-7454
FAX: (602) 451-9495

Interesting selection of surplus and liquidation items, as well as a good line of standard components, hardware, and kits. Here again, you never know what "goodies" you are going to find until you look through their catalog.

JAMECO ELECTRONIC COMPONENTS & COMPUTER PRODUCTS

1355 Shoreway Road
Belmont, CA 94002-4100
Order Line: (800) 831-4242
Customer Service: (415) 592-8097
FAX: (415) 592-2503

Complete selection of ICs, parts, tools, and equipment, with some of the best prices available. Also, has one of the better over-all selections of computer boards and accessories for both current and older machines. Nice choice of EPROM/EEPROM programmers, too.

JDR MICRODEVICES

1850 South 10th Street
San Jose, CA 95112-4108
Order Line: (800) 538-5000
Tech. Support: (800) 538-5002
Local: (408) 494-1400

FAX: (408) 494-1430

Excellent line of ICs, discrete components, kits, and equipment. Stocks several EPROM/EEPROM programmers, has good prices, and is highly customer service oriented. Frequent and separate catalogs for both electronic components and PC products.

MARLIN P. JONES & ASSOCIATES, INC.

P.O. Box 12685
Lake Park, FL 33403-0685
Order Line: (800) OK 2 ORDER (652-67337)
Customer Service: (407) 848-8236
FAX: (800) 4 FAX YES (432-9937)

Large selection of surplus and liquidation parts. Specializes in power supplies, motors, tools, and hardware, although they do carry kits, electronic equipment, and components.

MENDELSON ELECTRONICS, INC. (MECI)

340 East First Street
Dayton, OH 45402
Order Line: (800) 344-4465
Local: (513) 461-3525
FAX: (800) 344-6324
E-mail: meci@meci.com
Internet: http://www.meci.com

Carries good line of components, hardware, motors, cable, and connectors. Very few semiconductors, but does carry some interesting computer and general electronic surplus items.

MOUSER ELECTRONICS

National Circulation Center
2401 Highway 287 North
Mansfield, TX 76063-4827
Cust. Service/Order Line: (800) 346-6873
E-mail: sales@mouser.com
Internet: http://www.mouser.com

Integrated Circuit Projects

Huge selection of semiconductors, discrete components, hardware, switches, connectors, tools, cases and cabinets, replacement parts, etc. Good prices, and frequently revised catalogs.

NATIONAL SEMICONDUCTOR

P.O. Box 58090
Santa Clara, CA 95052-8090
Customer Service: (408) 721-5000
Tech. Support: (404) 564-5699
BBS: (408) 245-0671

A leader in semiconductor products, including LED displays and support integrated circuits. NS is a possible source of various seven segment LED units suitable for some of the projects.

RADIO SHACK

Tandy Corporation
Fort Worth, TX 76102
Local: (817) 390-3011

Stores nation-wide with a limited inventory of parts, hardware, and tools. Good variety of switches, knobs, and project boxes.

As stated at the outset, these are just some of the companies out there to help you. In each case, as many telephone numbers and other forms of communication as possible have been included, and as of this writing, these are correct. However, the phone company is handy at changing things, so you may have to update this information from time to time.

Also, many of these companies have a minimum order dollar limit, which you need to check on if you order by phone. As far as shipping/handling charges, they vary from company to company, with some charging flat rates, while others charge actual rates. Again, check on this if you call. However, it is recommended that you obtain a copy of each company's catalog or sales literature, as they will come in handy in the future.

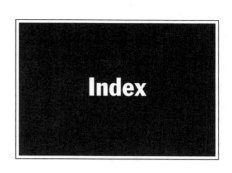

Index

A

A/D Converter 30, 37, 114, 115, 116, 117, 120
Address Bus 30
Addressing 33
AM 192
Ambient Light 189
Ambient Temperature 44, 122
Amperage 145, 146, 149
Amplifier 18, 37, 144, 170, 192, 193
Analog Meter 14, 18
Analog Storage 28
Anode 17, 25, 67, 115, 116, 149, 150
Anode Display 60, 64
Answering Machine 29
Antenna 111, 166
Anti-Drift 120
Audio Amplifier 37, 109
Audio Noise 33
Audio Range 36
Audio Section 33
Audio Transformer 17

B

Bandpass Filter 143, 144, 186
Bargraph Display Driver 102
Bargraph Driver 15
Baseband 166

Battery 110
BCD 24, 60, 65, 70, 74, 114, 122
BCD Code 21
BCD Device 51, 52
Beam Spread 154
Bell Laboratory 18, 61, 148
Bias 93, 150
Binary Code 21, 51
Binary Decoder 50
Binary Encoder 50
Bits 21, 30
Bridges 122
Bug Detector 18, 30, 100, 101, 102, 103, 104, 105
By-Pass 93
Byte 117

C

Capacitance 165
Capacitor 37, 53, 60, 109, 120, 121, 122, 123, 128, 129, 143, 151, 157, 158, 159, 163, 164, 165, 166, 176, 177, 192, 193, 194
Carrier 187
Carry Line 60
Cascade 30, 53, 59, 81, 86
Cathode 17, 136, 149
Center Tap 17
Channel 14

Integrated Circuit Projects

Charge Pump 174
Charge/Discharge Network 157
Chips 15
Clock 50, 54, 120, 170
CMOS 80, 82, 139
Code 23
Coherency 154
Coil 93, 95, 97, 109, 144,
 145, 146, 164, 165,
 166, 192, 194, 195
Collector 134, 135, 150
Collimation 154, 189
Color Burst Crystal 24, 73
Color Organ 144
Communications Transceiver 14
Comparator 44, 128, 174, 187
Comparator Circuit 15, 144
Complementary Metal Oxide
 Semiconductor 80
Control Logic 28
Converter 128
Counter 50, 53, 58, 59, 60,
 65, 69, 74, 82, 85,
 115, 122, 144
Counter Register 65
Counter/Display Driver 64
Counting System 64
Crystal 73, 121, 170
Crystal Oscillator 164
Current 150

D

D/A converter 30, 38
Data 29
Data Sheet 32, 154, 174
Decade Counter 51
Decimal Digit 51
Decimal Point 117, 124
Decoder 120, 171
Decoder Chip 21
Decoder/Driver 21, 24, 74,
 114, 115, 122
Detector 102, 103, 172,
 183, 185, 193
Digit Decoder 37
Digit Driver 37
Digital Counter 54

Digital Meter 39
Digital Recorder 28
Diode 60, 109, 120, 121,
 135, 137, 144, 149,
 151, 152, 153, 154,
 164, 165, 170, 194
Diode Laser 149
Diode Network 163
DIP 120
Disk Capacitor 95
Display 24, 25, 50, 52, 60,
 67, 74, 75, 77, 84,
 116, 117, 120, 122, 125
Display Driver 37, 50, 52
Divergence 154
Divider 82
Driver 50, 65, 66, 67,
 117, 120, 153
Driver Circuit 149, 152
Dropping Resistor 15
DTMF 168, 169, 171, 172
Dual Timer Chip 158

E

Earphones 109
EEPROM 29, 30
Einstein, Albert 148
Electronic Memo Pad 28
Emitter 91, 134, 135, 150
Encoder/Decoder Chip 185
Energy 151, 183
Epoxy 47
Error 120

F

Federal Communications
 Commission 95
FET 93
Field Effect Transistor 93
Field Strength Meter 112
Filter 28, 29, 143, 144
FM Modulator 164
Frequency 40, 85, 93, 182
Frequency By-Pass 93
Frequency Counter 70
Frequency Divider 70, 85
Frequency Meter 36, 39, 40, 131

Index

Function Generator 128

G

Gain 37, 93, 101, 103, 109, 116, 129, 164, 193
Gain Control 28, 33, 117
Gallium/Arsenide Junction 153
Gate 67, 150
Generator 130
Governor 174
Ground 32, 69, 109, 117, 124, 135, 139, 144, 163
Ground Plane 33
Ground Rail 123

H

Harris-Intersil 64
Headphones 111
Hex Digit Driver 75
Hex Inverter 65

I

Inductance 165
Induction System 109
Infrared Light 152, 181, 182, 183, 184, 185, 188
Input Transducer 122
Input Voltage 15, 16, 122
Intensity 14, 182
Interference 33
Intervalometer 157
Inverter 66, 136, 138, 139, 187
Inverter Gate 67
Inverting Potential 44
ISD Series 33

J

Johnson Counter 80, 87

L

Lamp 174
Laser 61, 148, 149, 150, 152, 153, 154
Laser Diode 148, 149, 150, 151, 152, 153, 154
Layout 68, 69, 76

LCD 74
LED 14, 15, 16, 18, 20, 21, 37, 50, 52, 64, 65, 67, 74, 75, 82, 84, 85, 102, 103, 115, 116, 122, 134, 135, 136, 137, 138, 150, 176, 177, 178, 184, 188, 189, 190
Light 181, 182, 183, 184
Light Wave 182
Load 109, 145, 146, 174
Logic 33, 134
Logic Chip 69
Logic Probe 134, 139, 140
Logic Scheme 66
Logic State 135

M

Megabit 30
Memory 30
Memory Cell 30
Metal Oxide Varistor 170
Meter 17, 39, 120, 121, 122, 124, 125, 174, 177
Microphone 18, 33, 36, 37, 40, 41, 93, 96, 103, 144, 189
Microphotography 61
Microprocessor 137, 140, 189
Modem 111
Modulator 163, 164, 166
Monitor 114
Morse Code 111
Mostek Corporation 72
MOV 170
Multimeter 114, 125
Multivibrator 40, 137, 139
Mylar 176

N

NAND Gate 65, 66
National Semiconductor 15
Noise 21, 33, 39
NPN 91, 93, 95, 115
Nybble 21

Integrated Circuit Projects

O

Op-Amp 44, 45, 46, 101, 102, 103, 143, 144, 145, 146, 149, 174, 175, 185, 186
Optical Collimator 189
Oscillation 128
Oscillator 24, 37, 39, 40, 65, 73, 75, 82, 85, 91, 92, 93, 94, 95, 97, 101, 103, 105, 108, 112, 121, 128, 157, 164, 166, 169
Oscilloscope 39
Output 85
Output Buffer 128

P

PA Amplifiers 14
Passive Detector 100, 102, 108, 109
PCB 17, 60, 68, 76, 116
Persistence of Vision 83
Photodiode 149, 150, 186
Phototransistor 184, 183, 185, 186, 189
Pickups 96, 176
Pinouts 15
PNP 91, 92, 93, 95
Point-to-Point Wiring 68
Positive Rail 58, 69, 139, 150, 163
Potentiometer 17, 37, 39, 40, 44, 51, 53, 82, 102, 103, 109, 111, 116, 122, 124, 125, 129, 144, 146, 150, 153, 157, 158, 159, 164, 171, 194
Power 67, 69, 111
Power Source 125, 130, 152
Power Supply 25, 37, 38, 53, 60, 97, 110, 116, 120, 122, 146, 152, 153, 154, 158, 170, 194
Power/Signal Meter 18
Pre-Amp Stage 144
Pre-Amplifier 28, 37, 93, 96, 109, 189
Probe 122, 136, 137, 139
Prototype 134
Pulse 50, 51, 55, 66, 120, 137, 148, 183, 184, 187

Q

Quad Segment Driver 75

R

Radiation Theory 148
Radio 108, 192
Radio Shack 32
Radio Wave 182
Random Number Generator 85
Receiver 14, 24, 101, 103, 108, 109, 111, 183, 184, 185, 188, 189, 192, 195, 196
Rectifier 185
Rectifier Bridge 170
Reference 120
Reference Voltage 15, 121
Reflector 189
Register 65, 66
Regulator 67, 116, 123, 124
Relay 144, 145, 146, 157, 174, 176, 183, 184, 185
Remote Control 184
Repeater 172
Resistor 15, 44, 52, 54, 60, 101, 117, 121, 124, 134, 143, 145, 151, 157, 164, 164, 165, 193
RF 102, 103, 183
RF Amplifier 193
RF Circuit 164
RF Energy 103, 182
RF Modulator 164
RF Oscillator 164
RF Receiver 100
RF Signal 101
RF Tank 166

S

Sample Rate 121
Sawtooth Signal 54

Index

Scanner 149
Schmitt Trigger 65, 187
Semiconductor 61, 91, 92
Serial Signal 183
Signal 21, 24
Signal Conditioning 24
Sine Wave 128, 129, 186
Solenoid 176
Sound Meter 40
SPDT 125
Speaker 40, 75, 109
Speaker Output 33
Speedometer 174
Spike 24, 144, 152
Spinel 148
Square Wave 54, 128, 129, 186
Square Wave Generator 82
Stopwatch 65, 68
Storage 29
Superheterodyne 108
Supply Rail 17
Supply Voltage 15, 145
Surge 152, 153
Switch 58

T

Tachometer 174, 175, 176, 177, 178
Tank Circuit 164
Tape 29
Tape Deck 14
Telephone 168, 169, 170, 171, 172
Telex Signal 111
Temperature 44, 45, 122
Temperature Detector 124
Temperature Meter 125
Temperature Monitor 46
Temperature Probe 122, 123
Thermistor 44, 45, 46, 122, 125
Thermometer 114, 120, 125
Threshold 154
Timer 50, 51, 54, 82, 169
Timing Unit 65
Tone 146, 157, 159, 169
Tone Decoder 21, 23, 24
Tone Generator 158, 168, 169

Tone Pair 20, 21, 25
Touch-Tone System 168
Transceiver 112
Transducer 41, 122
Transformer 21, 24, 95, 109, 122, 123, 144
Transient Signal 123, 153
Transient 24
Transistor 37, 60, 91, 92, 93, 109, 115, 134, 149, 150, 151, 184, 185, 187, 192, 193
Transistor Amplifier 75
Transitory Energy 151
Transmitter 91, 93, 95, 101, 102, 103, 111, 112, 114, 163, 183, 184, 185, 188, 189
Triangle Wave 128, 129
TTL 58, 60, 65, 139
TTL Logic 139
TTL Series 53, 58
Tuning Fork Crystal 73

U

Ultrasonic Transducer 41

V

Variable Coil 95
Varistor 21, 24
Video 166
Video Monitor 166
Video/Audio Input 166
Visible Light 151
VLF 196
Voice Link 189
Voltage 114, 116, 120, 121, 130, 139, 144, 154
Voltage Amp 37
Voltage Divider 15, 37, 44, 93, 101, 122
Voltage Meter 37
Voltage Rail 93
Voltage Regulators 120, 122
Voltmeter 37, 154, 178
Volume 109, 111, 171, 185, 194
Volume Control 144

Integrated Circuit Projects

VU Meter 14, 18

W

Wattage 145
Waveform 128, 129
Wave 128, 129, 182
Wiring Scheme 129, 158

X

Xenon 148

Z

Zener Diode 25, 121

Howard W. Sams
A Bell Atlantic Company

Your Technology Connection to the Future!

Now You Can Visit Howard W. Sams & Company <u>On-Line</u>: http://www.hwsams.com

Gain Easy Access to:

- **The PROMPT Publications catalog, for information on our *Latest Book Releases*.**
- **The PHOTOFACT Annual Index.**
- **Information on Howard W. Sams' Latest Products.**
- ***AND MORE!***

PROMPT
PUBLICATIONS

A Division of Howard W. Sams & Company
A Bell Atlantic Company

CALL 1-800-428-7267 TODAY FOR THE NAME OF YOUR NEAREST PROMPT PUBLICATIONS DISTRIBUTOR

PC Hardware Projects Volume 1
James "J.J." Barbarello

Real-World Interfacing with Your PC
James "J.J." Barbarello

Now you can create your own PC-based digital design workstation! Using commonly available components and standard construction techniques, you can build some key tools to troubleshoot digital circuits and test your printer, fax, modem, and other multi-conductor cables.

This book will guide you through the construction of a channel logic analyzer, and a multipath continuity tester. You will also be able to combine the projects with an appropriate power supply and a prototyping solderless breadboard system into a single digital workstation interface!

As the computer becomes increasingly prevalent in society, its functions and applications continue to expand. Modern software allows users to do everything from balance a checkbook to create a family tree. Interfacing, however, is truly the wave of the future for those who want to use their computer for things other than manipulating text, data, and graphics.

Real-World Interfacing With Your PC provides all the information necessary to use a PC's parallel port as a gateway to electronic interfacing.

Computer Technology
256 pages ♦ Paperback ♦
7-3/8 x 9-1/4"
ISBN: 0-7906-1104-X ♦ Sams:
61104
$24.95 ♦ Feb. 1997

Computer Technology
119 pages ♦ Paperback ♦
7-3/8 x 9-1/4"
ISBN: 0-7906-1078-7 ♦ Sams:
61078
$16.95 ♦ March 1996

CALL 1-800-428-7267 TODAY FOR THE NAME OF YOUR NEAREST PROMPT PUBLICATIONS DISTRIBUTOR

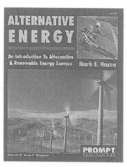

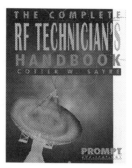

Alternative Energy
Mark E. Hazen

The Complete RF Technician's Handbook
Cotter W. Sayre

This book is designed to introduce readers to the many different forms of energy mankind has learned to put to use. Generally, energy sources are harnessed for the purpose of producing electricity. This process relies on transducers to transform energy from one form into another. *Alternative Energy* will not only address transducers and the five most common sources of energy that can be converted to electricity, it will also explore solar energy, the harnessing of the wind for energy, geothermal energy, and nuclear energy.

The *Complete RF Technician's Handbook* will furnish the working technician or student with a solid grounding in the latest methods and circuits employed in today's RF communications gear. It will also give readers the ability to test and troubleshoot transmitters, transceivers, and receivers with absolute confidence. Some of the topics covered include reactance, phase angle, logarithms, diodes, passive filters, amplifiers, and distortion. Various multiplexing methods and data, satellite, spread spectrum, cellular, and microwave communication technologies are discussed.

Professional Reference
320 pages ◆ Paperback ◆
7-3/8 x 9-1/4"
ISBN: 0-7906-1079-5 ◆ Sams: 61079
$18.95 ($25.95 Canada) ◆ October 1996

Professional Reference
281 pages ◆ Paperback ◆
8-1/2 x 11"
ISBN: 0-7906-1085-X ◆ Sams: 61085
$24.95 ($33.95 Canada) ◆ July 1996

**CALL 1-800-428-7267 TODAY FOR THE NAME OF
YOUR NEAREST PROMPT PUBLICATIONS DISTRIBUTOR**

ES&T Presents TV Troubleshooting & Repair
Electronic Servicing & Technology Magazine

ES&T Presents Computer Troubleshooting & Repair
Electronic Servicing & Technology

TV set servicing has never been easy. The service manager, service technician, and electronics hobbyist need timely, insightful information in order to locate the correct service literature, make a quick diagnosis, obtain the correct replacement components, complete the repair, and get the TV back to the owner.

ES&T Presents TV Troubleshooting & Repair presents information that will make it possible for technicians and electronics hobbyists to service TVs faster, more efficiently, and more economically.

ES&T is the nation's most popular magazine for professionals who service consumer electronics equipment. PROMPT® Publications, a rising star in the technical publishing business, is combining its publishing expertise with the experience and knowledge of *ES&T's* best writers to produce a new line of troubleshooting and repair books for the electronics market. Compiled from articles and prefaced by the editor in chief, Nils Conrad Persson, these books provide valuable, hands-on information for anyone interested in electronics and product repair.

Video Technology
226 pages ♦ Paperback ♦ 6 x 9"
ISBN: 0-7906-1086-8 ♦ Sams: 61086
$18.95 ($25.95 Canada) ♦ August 1996

Computer Technology
288 pages ♦ Paperback ♦ 6 x 9"
ISBN: 0-7906-1087-6 ♦ Sams: 61087
$18.95 ($26.50 Canada) ♦ February 1997

CALL 1-800-428-7267 TODAY FOR THE NAME OF YOUR NEAREST PROMPT PUBLICATIONS DISTRIBUTOR

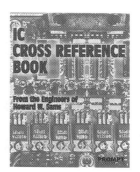

Semiconductor Cross Reference Book Fourth Edition
Howard W. Sams & Company

IC Cross Reference Book Second Edition
Howard W. Sams & Company

This newly revised and updated reference book is the most comprehensive guide to replacement data available for engineers, technicians, and those who work with semiconductors. With more than 490,000 part numbers, type numbers, and other identifying numbers listed, technicians will have no problem locating the replacement or substitution information needed. There is not another book on the market that can rival the breadth and reliability of information available in the fourth edition of the *Semiconductor Cross Reference Book*.

The engineering staff of Howard W. Sams & Company assembled the *IC Cross Reference Book* to help readers find replacements or substitutions for more than 35,000 ICs and modules. It is an easy-to-use cross reference guide and includes part numbers for the United States, Europe, and the Far East. This reference book was compiled from manufacturers' data and from the analysis of consumer electronics devices for PHOTOFACT® service data, which has been relied upon since 1946 by service technicians worldwide.

Professional Reference
688 pages ♦ Paperback ♦
8-1/2 x 11"
ISBN: 0-7906-1080-9 ♦ Sams: 61080
$24.95 ($33.95 Canada) ♦ August 1996

Professional Reference
192 pages ♦ Paperback ♦
8-1/2 x 11"
ISBN: 0-7906-1096-5 ♦ Sams: 61096
$19.95 ($26.99 Canada) ♦ November 1996

CALL 1-800-428-7267 TODAY FOR THE NAME OF YOUR NEAREST PROMPT PUBLICATIONS DISTRIBUTOR

The Component Identifier and Source Book
Victor Meeldijk

Tube Substitution Handbook
William Smith & Barry Buchanan

Because interface designs are often reverse engineered using component data or block diagrams that list only part numbers, technicians are often forced to search for replacement parts armed only with manufacturer logos and part numbers.

This source book was written to assist technicians and system designers in identifying components from prefixes and logos, as well as find sources for various types of microcircuits and other components. There is not another book on the market that lists as many manufacturers of such diverse electronic components.

The most accurate, up-to-date guide available, the *Tube Substitution Handbook* is useful to antique radio buffs, old car enthusiasts, and collectors of vintage ham radio equipment. In addition, marine operators, microwave repair technicians, and TV and radio technicians will find the *Handbook* to be an invaluable reference tool.

The *Tube Substitution Handbook* is divided into three sections, each preceded by specific instructions. These sections are vacuum tubes, picture tubes, and tube basing diagrams.

Professional Reference
384 pages ◆ Paperback ◆ 8-1/2 x 11"
ISBN: 0-7906-1088-4 ◆ Sams: 61088
$24.95 ($33.95 Canada) ◆ November 1996

Professional Reference
149 pages ◆ Paperback ◆ 6 x 9"
ISBN: 0-7906-1036-1 ◆ Sams: 61036
$16.95 ($22.99 Canada) ◆ March 1995

CALL 1-800-428-7267 TODAY FOR THE NAME OF YOUR NEAREST PROMPT PUBLICATIONS DISTRIBUTOR

Speakers for Your Home & Automobile
Gordon McComb, Alvis J. Evans, & Eric J. Evans

Sound Systems for Your Automobile
Alvis J. Evans & Eric J. Evans

The cleanest CD sound, the quietest turntable, or the clearest FM signal are useless without a fine speaker system. This book not only tells readers how to build quality speaker systems, it also shows them what components to choose and why. The comprehensive coverage includes speakers, finishing touches, construction techniques, wiring speakers, and automotive sound systems.

This book provides the average vehicle owner with the information and skills needed to install, upgrade, and design automotive sound systems. From terms and definitions straight up to performance objectives and cutting layouts, *Sound Systems* will show the reader how to build automotive sound systems that provide occupants with live performance reproductions that rival home audio systems.

Whether starting from scratch or upgrading, this book uses easy-to-follow steps to help readers plan their system, choose components and speakers, and install and interconnect them to achieve the best sound quality possible.

Audio Technology
164 pages ◆ Paperback ◆ 6 x 9"
ISBN: 0-7906-1025-6 ◆ Sams: 61025
$14.95 ($20.95 Canada) ◆ November 1992

Audio Technology
124 pages ◆ Paperback ◆ 6 x 9"
ISBN: 0-7906-1046-9 ◆ Sams: 61046
$16.95 ($22.99 Canada) ◆ January 1994

CALL 1-800-428-7267 TODAY FOR THE NAME OF YOUR NEAREST PROMPT PUBLICATIONS DISTRIBUTOR

Is This Thing On?	**Advanced Speaker Designs**
Gordon McComb	*Ray Alden*

Is This Thing On? takes readers through each step of selecting components, installing, adjusting, and maintaining a sound system for small meeting rooms, churches, lecture halls, public-address systems for schools or offices, or any other large room.

In easy-to-understand terms, drawings and illustrations, *Is This Thing On?* explains the exact procedures behind connections and troubleshooting diagnostics. With the help of this book, hobbyists and technicians can avoid problems that often occur while setting up sound systems for events and lectures.

Advanced Speaker Designs shows the hobbyist and the experienced technician how to create high-quality speaker systems for the home, office, or auditorium. Every part of the system is covered in detail, from the driver and crossover network to the enclosure itself. Readers can build speaker systems from the parts lists and instructions provided, or they can actually learn to calculate design parameters, system responses, and component values with scientific calculators or PC software.

Audio Technology
136 pages ♦ Paperback ♦ 6 x 9"
ISBN: 0-7906-1081-7 ♦ Sams: 61081
$14.95 ($20.95 Canada) ♦ April 1996

Audio Technology
136 pages ♦ Paperback ♦ 6 x 9"
ISBN: 0-7906-1070-1 ♦ Sams: 61070
$16.95 ($22.99 Canada) ♦ July 1995

CALL 1-800-428-7267 TODAY FOR THE NAME OF YOUR NEAREST PROMPT PUBLICATIONS DISTRIBUTOR

Theory & Design of Loudspeaker Enclosures
Dr. J. Ernest Benson

Making Sense of Sound
Alvis J. Evans

The design of loudspeaker enclosures, particularly vented enclosures, has been a subject of continuing interest since 1930. Since that time, a wide range of interests surrounding loudspeaker enclosures have sprung up that grapple with the various aspects of the subject, especially design. *Theory & Design of Loudspeaker Enclosures* lays the groundwork for readers who want to understand the general functions of loudspeaker enclosure systems and eventually experiment with their own design.

This book deals with the subject of sound — how it is detected and processed using electronics in equipment that spans the full spectrum of consumer electronics. It concentrates on explaining basic concepts and fundamentals to provide easy-to-understand information, yet it contains enough detail to be of high interest to the serious practitioner. Discussion begins with how sound propagates and common sound characteristics, before moving on to the more advanced concepts of amplification and distortion. *Making Sense of Sound* was designed to cover a broad scope, yet in enough detail to be a useful reference for readers at every level.

Audio Technology
244 pages ♦ Paperback ♦ 6 x 9"
ISBN: 0-7906-1093-0 ♦ Sams: 61093
$19.95 ($26.99 Canada) ♦ August 1996

Audio Technology
112 pages ♦ Paperback ♦ 6 x 9"
ISBN: 0-7906-1026-4 ♦ Sams: 61026
$10.95 ($14.95 Canada) ♦ November 1992

**CALL 1-800-428-7267 TODAY FOR THE NAME OF
YOUR NEAREST PROMPT PUBLICATIONS DISTRIBUTOR**

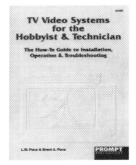

The Video Book
Gordon McComb

TV Video Systems
L.W. Pena & Brent A. Pena

Televisions and video cassette recorders have become part of everyday life, but few people know how to get the most out of these home entertainment devices. *The Video Book* offers easy-to-read text and clearly illustrated examples to guide readers through the use, installation, connection, and care of video system components. Simple enough for the new buyer, yet detailed enough to assure proper connection of the units after purchase, this book is a necessary addition to the library of every modern video consumer.

Knowing which video programming source to choose, and knowing what to do with it once you have it, can seem overwhelming. Covering standard hard-wired cable, large-dish satellite systems, and DSS, *TV Video Systems* explains the different systems, how they are installed, their advantages and disadvantages, and how to troubleshoot problems. This book presents easy-to-understand information and illustrations covering installation instructions, home options, apartment options, detecting and repairing problems, and more. The in-depth chapters guide you through your TV video project to a successful conclusion.

Video Technology
192 pages ♦ Paperback ♦ 6 x 9"
ISBN: 0-7906-1030-2 ♦ Sams: 61030
$16.95 ($22.99 Canada) ♦ October 1992

Video Technology
124 pages ♦ Paperback ♦ 6 x 9"
ISBN: 0-7906-1082-5 ♦ Sams: 61082
$14.95 ($20.95 Canada) ♦ June 1996

**CALL 1-800-428-7267 TODAY FOR THE NAME OF
YOUR NEAREST PROMPT PUBLICATIONS DISTRIBUTOR**

The Howard W. Sams Troubleshooting & Repair Guide to TV
Howard W. Sams & Company

The In-Home VCR Mechanical Repair & Cleaning Guide
Curt Reeder

The Howard W. Sams Troubleshooting & Repair Guide to TV is the most complete and up-to-date television repair book available. Included in its more than 300 pages is complete repair information for all makes of TVs, time-saving features that even the pros don't know, comprehensive basic electronics information, and extensive coverage of common TV symptoms.

This repair guide is completely illustrated with useful photos, schematics, graphs, and flowcharts. It covers audio, video, technician safety, test equipment, power supplies, picture-in-picture, and much more.

Like any machine that is used in the home or office, a VCR requires minimal service to keep it functioning well and for a long time. However, a technical or electrical engineering degree is not required to begin regular maintenance on a VCR. *The In-Home VCR Mechanical Repair & Cleaning Guide* shows readers the tricks and secrets of VCR maintenance using just a few small hand tools, such as tweezers and a power screwdriver.

This book is also geared toward entrepreneurs who may consider starting a new VCR service business of their own.

Video Technology
384 pages Paperback ♦
8-1/2 x 11"
ISBN: 0-7906-1077-9 ♦ Sams: 61077
$29.95 ($39.95 Canada) ♦ June 1996

Video Technology
222 pages ♦ Paperback ♦
8-3/8 x 10-7/8"
ISBN: 0-7906-1076-0 ♦ Sams: 61076
$19.95 ($26.99 Canada) ♦ April 1996

CALL 1-800-428-7267 TODAY FOR THE NAME OF YOUR NEAREST PROMPT PUBLICATIONS DISTRIBUTOR

Surface-Mount Technology for PC Boards
James K. Hollomon, Jr.

Digital Electronics
Stephen Kamichik

The race to adopt surface-mount technology, or SMT as it is known, has been described as the latest revolution in electronics. This book is intended for the working engineer or manager, the student or the interested layman, who would like to learn to deal effectively with the many trade-offs required to produce high manufacturing yields, low test costs, and manufacturable designs using SMT. The valuable information presented in *Surface-Mount Technology for PC Boards* includes the benefits and limitations of SMT, SMT and FPT components, manufacturing methods, reliability and quality assurance, and practical applications.

Professional Reference
510 pages ◆ Paperback ◆ 7 x 10"
ISBN: 0-7906-1060-4 ◆ Sams: 61060
$26.95 ($36.95 Canada) ◆ July 1995

Although the field of digital electronics emerged years ago, there has never been a definitive guide to its theories, principles, and practices — until now. *Digital Electronics* is written as a textbook for a first course in digital electronics, but its applications are varied.

Useful as a guide for independent study, the book also serves as a review for practicing technicians and engineers. And because *Digital Electronics* does not assume prior knowledge of the field, the hobbyist can gain insight about digital electronics.

Electronic Theory
150 pages ◆ Paperback ◆
7-3/8 x 9-1/4"
ISBN: 0-7906-1075-2 ◆ Sams: 61075
$16.95 ($22.99 Canada) ◆ February 1996

**CALL 1-800-428-7267 TODAY FOR THE NAME OF
YOUR NEAREST PROMPT PUBLICATIONS DISTRIBUTOR**